Medicinal Plants of Ecuador

This unique volume draws on the rich culture, folklore and environment of medicinal plants in Ecuador, which includes the important rain forest region. This country has rich plant resources and a large diversity of plants. In particular, the Biotechnological Research Center of Ecuador, CIBE, is an important center of plant research and biodiversity. For more than 16 years, CIBE has been performing scientific research on plants and microorganisms with extensive focus on biodiversity, biotechnology, bioproducts, hytopathology, molecular biology, tissue culture and technology transfer. The Center has state-of-the-art infrastructure and technology and great strength in human resources.

Natural Products Chemistry of Global Plants

Editor: Raymond Cooper

This unique book series focuses on the natural products chemistry of botanical medicines from different countries such as Sri Lanka, Cambodia, Brazil, China, Africa, Borneo, Thailand, and Silk Road Countries. These fascinating volumes are written by experts from their respective countries. The series will focus on the pharmacognosy, covering recognized areas rich in folklore as well as botanical medicinal uses as a platform to present the natural products and organic chemistry. Where possible, the authors will link these molecules to pharmacological modes of action. The series intends to trace a route through history from ancient civilizations to the modern day showing the importance to man of natural products in medicines, foods, and a variety of other ways.

Recent Titles in this Series:

Natural Products Chemistry of Botanical Medicines from Cameroonian Plants
Xavier Siwe Noundou

Medicinal Plants and Mushrooms of Yunnan Province of China
Clara Lau and Chun-Lin Long

Medicinal Plants of Borneo
Simon Gibbons and Stephen P. Teo

Natural Products and Botanical Medicines of Iran
Reza Eddin Owfi

Natural Products of Silk Road Plants
Raymond Cooper and Jeffrey John Deakin

Brazilian Medicinal Plants
Luzia Modolo and Mary Ann Foglio

Medicinal Plants of Bangladesh and West Bengal: Botany, Natural Products, and Ethnopharmacology
Christophe Wiart

Traditional Herbal Remedies of Sri Lanka
Viduranga Y. Waisundara

Medicinal Plants of Ecuador
Pablo Chong Aguirre, Migdalia Miranda Martínez, Patricia Manzano Santana (Eds)

Medicinal Plants of Ecuador

Edited by
Pablo Chong Aguirre
Escuela Superior Politécnica del Litoral, Ecuador

Migdalia Miranda Martínez
ESPOL, Facultad de Ciencias Naturales y Matemáticas, Ecuador

Patricia Manzano Santana
ESPOL, Facultad de Ciencias de la Vida, Ecuador

CRC Press is an imprint of the
Taylor & Francis Group, an **informa** business

First edition published 2023
by CRC Press
6000 Broken Sound Parkway NW, Suite 300, Boca Raton, FL 33487–2742

and by CRC Press
4 Park Square, Milton Park, Abingdon, Oxon, OX14 4RN

CRC Press is an imprint of Taylor & Francis Group, LLC

© 2023 selection and editorial matter, Pablo Chong Aguirre, Patricia Manzano Santana, Migdalia Miranda Martínez; individual chapters, the contributors

Reasonable efforts have been made to publish reliable data and information, but the author and publisher cannot assume responsibility for the validity of all materials or the consequences of their use. The authors and publishers have attempted to trace the copyright holders of all material reproduced in this publication and apologize to copyright holders if permission to publish in this form has not been obtained. If any copyright material has not been acknowledged please write and let us know so we may rectify in any future reprint.

Except as permitted under U.S. Copyright Law, no part of this book may be reprinted, reproduced, transmitted, or utilized in any form by any electronic, mechanical, or other means, now known or hereafter invented, including photocopying, microfilming, and recording, or in any information storage or retrieval system, without written permission from the publishers.

For permission to photocopy or use material electronically from this work, access www.copyright.com or contact the Copyright Clearance Center, Inc. (CCC), 222 Rosewood Drive, Danvers, MA 01923, 978–750–8400. For works that are not available on CCC please contact mpkbookspermissions@tandf.co.uk

Trademark notice: Product or corporate names may be trademarks or registered trademarks and are used only for identification and explanation without intent to infringe.

ISBN: 978-0-367-77586-5 (HB)
ISBN: 978-1-032-00398-6 (PB)
ISBN: 978-1-003-17399-1 (EB)

DOI: 10.1201/9781003173991

Typeset in Times
by Apex CoVantage, LLC

Pablo Chong Aguirre would like to dedicate this book to his family, Adriana de Chong, Abigail Chong and Emily Chong, and to his colleagues, Migdalia Miranda, Patricia Manzano and all of the CIBE-ESPOL team.

In memory of Migdalia Miranda Martínez

"A wonderful human being and an extraordinary Cuban scientist and academic. Migdalia was a pillar in our community and an example to follow. She gave her heart and soul to her work and friends and was always a helping hand and support for her colleagues. We´ll always miss her beautiful warm smile, her jovial and calm character, and her big heart."

Pablo Chong Aguirre

"To a great professor, Migdalia,

Exceptional and tireless worker; overflowing joy; spark and strength of character that radiated and spread in her walk. For her, aging was not a barrier nor a disenchantment with life; she fought and overcame adversity . . . she was an unconditional Mother; teacher-guide; jovial friend and life counselor . . . virtues impossible to achieve . . . but worthy of imitation. My true north . . . really a GREAT TEACHER, that fulfilled the designs of God. The Word of God is a Word without sound, but with efficiency. Word that is not perceived with the ear, but that is captured with the heart. Word that is not shouted, but that is felt in the innermost intimacy of our being.

Migdalia, we will always remember you!"

Patricia Manzano Santana

Contents

Ecuador: An Introduction

Ecuador is a unique, beautiful and small country possessing a huge biodiversity of flora. Ecuador is located on the west coast of South America by the Pacific Ocean, bordering to the north with Colombia and to the south and east with Peru (see Figure 1). The strategic geographical location of this tiny country on the equator makes it one of the megadiverse hotspots in fauna and flora of the world.

Ecuador possesses a huge variety of regions and biomes with a long coast influenced by El Niño and Humboldt Ocean currents, the Highlands with some of the biggest Andean volcanos. It is part of the Amazon Jungle and home to the unique Galapagos Islands.

This diversity in biomes encompasses a huge potential for the discovery of native medicinal plants and new drugs to treat many important diseases, yet these discoveries may not be known to western civilization. For millennia native communities have had ancestral knowledge of the uses of native plants for treatment or cure of different

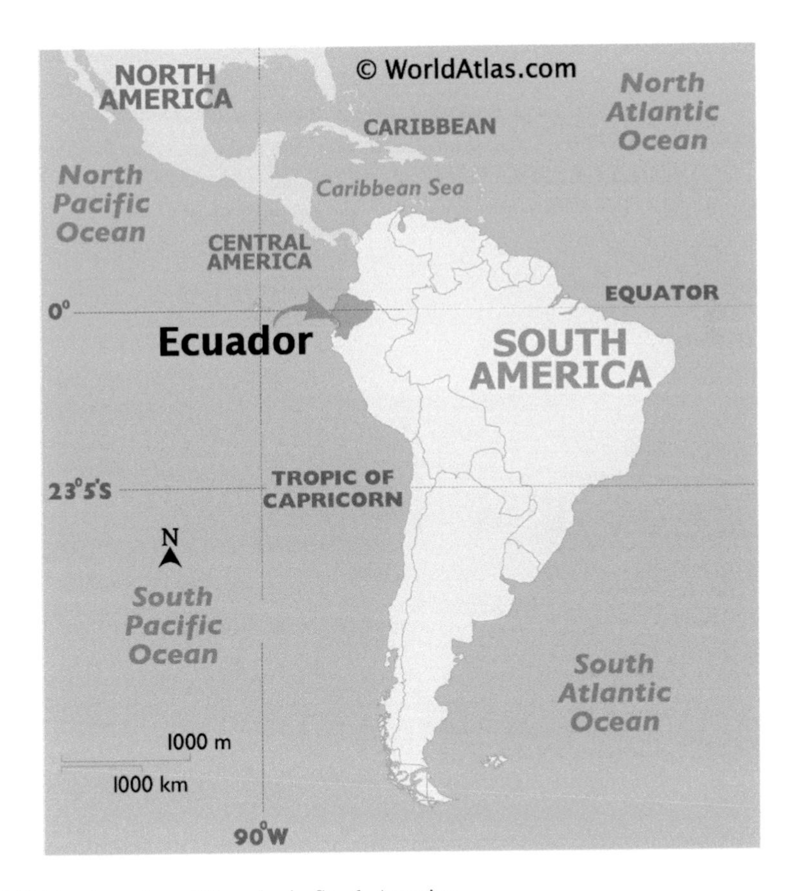

FIGURE 1 Location of Ecuador in South America.

diseases. However, with the relevant ongoing deforestation and fragmentation of the country terrain ecosystems it is of the utmost importance to save these species and their ancient knowledge.

This book, *Medicinal Plants of Ecuador*, attempts to provide knowledge that now can be shared with the international community. The aim of the book is to present the latest scientific research on native medicinal plants from Ecuador, contributing to an increase in the knowledge on human health, pharmacognosy, biodiversity, bioindustry and an attempt to encourage conservation.

The biodiversity of Ecuador encloses huge plant genetic and metabolic resources and invaluable knowledge for societal welfare. This book will allow this unique knowledge to be available to a broader audience. Since most Ecuadorian ancestral and scientific research has been kept hidden from the worldwide scientific community by linguistic barriers, this book permits this knowledge to be distributed and known. This book provides insights in the scientific bases of the traditional and ancient knowledge recovered from Ecuador natives and ancient cultures. Ancient knowledge that had and will guide us into the discovery of promising new drugs and treatments for human health.

Pablo Chong Aguirre, PhD
Molecular Biology Head Department
Centro de Investigaciones Biotecnológicas del Ecuador, CIBE
ESPOL Polytechnic University, Escuela Superior Politécnica del Litoral, ESPOL, Facultad de Ciencias de la Vida, Campus Gustavo Galindo, Km. 30.5 vía Perimetral, P.O. Box 09–01–5863, Guayaquil, Ecuador

Acknowledgments

Plant biodiversity is an invaluable natural resource for the discovery of new medicinal molecules. It is expected that pharmacognosy, in combination with other disciplines such as biotechnology, physiology, biochemistry, etc., provides extraordinary solutions for the treatment of numerous health problems and diseases. The integration of all these disciplines together with ancient knowledge will make possible the discovery and development of new drugs and better treatments for patients all over the world. With the application of these technologies, we can expand the products obtained from plant species, increasing economic and sustainability opportunities in the era of bioeconomy and human health.

We want to thanks to all the editors and authors of the chapters of this book for generously sharing the results of their research and discoveries. Thanks also to ESPOL University, Biotechnological Research Centre of Ecuador (CIBE), for creating the framework so that the authors of this book can work together in the area of drug discovery to provide the knowledge that ultimately will result in health and social benefits.

Editor Biographies

Pablo Chong Aguirre has 17 years of experience in biotechnology, molecular biology and plant pathology. He has a PhD degree in biotechnology obtained at Wageningen University and Research, Wageningen, the Netherlands. In 2001, Pablo started his research career as junior researcher in plant pathology at the Ecuadorian Biotechnology Research Centre from Escuela Superior Politécnica del Litoral, ESPOL University. In 2007 he obtained his master's degree in Biotechnology at ESPOL and became the molecular biology head in the ESPOL Biotechnology Research Centre until 2009 and the plant pathology head until 2013. In September 2012, he started his PhD program in the Netherlands at the Plant Research International banana group, and Laboratory of Phytopathology from Wageningen University and Research Centre on the analysis of fungicide resistance in *Pseudocercospora fijiensis* (Banana Black Sigatoka disease). His main research focused on fungi molecular resistance mechanisms to systemic fungicides. His research has led to publications related to a new resistance mechanism discovered in the species. Since 2016 he continues his work as invited Biology Professor of the Life Science Faculty at ESPOL University and Research Head Coordinator of the molecular biology and plant tissue culture departments at CIBE-ESPOL.

Migdalia Miranda Martínez was a professor and a scientific researcher with 41 years of experience in pharmacognosy and natural products chemistry. She has a PhD in chemical sciences since 1990 from the University of Havana, Cuba. She is a consulting professor at the University of Havana and a full-time professor-researcher hired from ESPOL. She holds the highest category of researcher granted by Senescyt (Principal Investigator 4) and was an advisor to the Prometheus Program of Senescyt for 3 years at the Center for Biotechnological Research CIBE-ESPOL. She has tutored 29 master theses and 8 PhDs. She has published 53 articles in indexed journals and more than 59 in other journals. She has also published 4 textbooks and 4 book chapters. Sadly, Migdalia passed away on March 21, 2021 at age 73 of Covid-19.

Patricia Manzano Santana has 30 years of experience in chemical and biological characterization of Ecuadorian biodiversity and the development of bioproducts useful for health, food, agriculture, construction and industry in general. She has a master's degree in pharmaceutical chemistry, a PhD in chemistry and pharmacy and another PhD in pharmaceutical sciences. She is a full-time professor at ESPOL. She is the research coordinator of the Bioproducts and Bioprocesses Laboratory of the Research Center of Ecuador (CIBE-ESPOL). She has tutored 22 undergraduate theses, 8 master's theses and 3 PhD theses. She has published 55 publications, 43 in indexed journals and 22 in Scopus in the last 5 years, including the publication of 4 chapters in books and the edition of 2 books. She has made 8 research stays, 3 post-doctoral.

Contributors

Manuel E. Baldeón
Facultad de Ciencias Médicas, de la Salud y de la Vida, Escuela de Medicina
Universidad Internacional del Ecuador
Quito, Ecuador

Zoraida Burbano
Facultad de Ciencias Químicas
Universidad de Guayaquil
Guayaquil, Ecuador

Elena Coyago Cruz
Carrera de Ingeniería en Biotecnología de los Recursos Naturales
Universidad Politécnica Salesiana, Sede Quito, Campus El Girón
Quito, Ecuador

Bárbara Beatriz Garrido Suarez
Departamento de Farmacia, Instituto de Farmacia y Alimentos (IFAL)
Universidad de La Habana
La Habana, Cuba

Alejandro Felipe González
Departamento de Farmacia, Instituto de Farmacia y Alimentos (IFAL)
Universidad de La Habana
La Habana, Cuba

Yamilet Irene Gutiérrez Gaitén
Department of Pharmacy, Institute of Pharmacy and Food
Universidad de La Habana
La Habana, Cuba

René Delgado Hernández
Centro de Investigaciones y Evaluaciones Biológicas, Instituto de Farmacia y Alimentos (IFAL)
Universidad de La Habana
La Habana, Cuba

Ramón Scull Lizama
Department of Pharmacy, Institute of Pharmacy and Food
Universidad de La Habana
La Habana, Cuba

Alexandra Jenny López Barrera
Faculty of Chemical Sciences
Universidad de Guayaquil
Guayaquil, Ecuador

Raisa Mangas
Instituto de Farmacia y Alimentos
Universidad de La Habana
La Habana, Cuba

Patricia Manzano Santana
Facultad de Ciencias de la Vida (FCV), Centro de Investigaciones Biotecnológicas del Ecuador (CIBE), Escuela Superior Politécnica del Litoral, Campus Gustavo Galindo
ESPOL Polytechnic University
Guayaquil, Ecuador

Migdalia Miranda Martínez
Facultad de Ciencias Naturales y Matemáticas, Campus Gustavo Galindo
Centro de Investigaciones Biotecnológicas del Ecuador (CIBE), Campus Gustavo Galindo
ESPOL Polytechnic University
Guayaquil, Ecuador

Carlos Ordoñez
Facultad de Ciencias de la Vida (FCV), Campus Gustavo Galindo
ESPOL Polytechnic University
Guayaquil, Ecuador

Andrea Orellana-Manzano
Facultad de Ciencias de la Vida (FCV), Laboratory for Biomedical Research, Campus Gustavo Galindo
ESPOL Polytechnic University
Guayaquil, Ecuador
Escuela de Odontología, Universidad Espíritu Santo
Guayaquil, Ecuador

Adriana Orellana-Paucar
Carrera de Nutrición y Dietética, Facultad de Ciencias Médicas
Universidad de Cuenca
Ciudadela El Paraíso s/n, Cuenca, Ecuador

Ricardo Pacheco
Centro de Investigaciones Biotecnológicas del Ecuador, Campus Gustavo Galindo
ESPOL Polytechnic University
Guayaquil, Ecuador

Juan Abreu Payrol
Entidad de Ciencia, Tecnologia e Innovacion Sierra Maestra, ECTI
Latin American School of Medicine
La Habana, Cuba

Juan Peñarreta
Facultad de Ciencias de la Vida, Campus Gustavo Galindo
ESPOL Polytechnic University
Guayaquil, Ecuador

Glenda Pilozo
Centro de Investigaciones Biotecnológicas del Ecuador (CIBE), Campus Gustavo Galindo
ESPOL Polytechnic University,
Guayaquil, Ecuador

Idania Rodeiro Guerra
Ministerio de Ciencia, Tecnología y Medio Ambiente
Instituto de Ciencias del Mar
La Habana, Cuba

Efrén Germán Santos Ordoñez
Facultad de Ciencias de la Vida, Campus Gustavo Galindo
Centro de Investigaciones Biotecnológicas del Ecuador, Campus Gustavo Galindo
ESPOL Polytechnic University,
Guayaquil, Ecuador

Glenda Sarmiento
Facultad de Ciencias Químicas
Universidad de Guayaquil
Guayaquil, Ecuador

Pilar Soledispa
Facultad de Ciencias Químicas
Universidad de Guayaquil
Guayaquil, Ecuador

Liliana Villao
Centro de Investigaciones Biotecnológicas del Ecuador, Campus Gustavo Galindo
ESPOL Polytechnic University
Guayaquil, Ecuador

Part I

Quality Control Considerations for South American Plants and Products

This section introduces the most common protocols and procedures for characterization, quality control and validation of herbal drugs and potential pharmaceutical products. This section also elaborates on the molecular identification of plant species base on DNA barcodes.

DOI: 10.1201/9781003173991-1

1 Importance of Pharmacognosy Studies in the Quality Control of Herbal Drugs

Yamilet Irene Gutiérrez Gaitén, Ramón Scull Lizama and Juan Abreu Payrol

CONTENTS

INTRODUCTION

Medicinal plants have traditionally been used in almost all cultures as a relevant therapeutic resource, since they serve as potential sources of biologically active compounds in the synthesis and development of new drugs (Jamshidi et al. 2018; Bhusnure et al. 2019). It is estimated that 80% of the world's population trusts herbal medicine and uses it in primary health care (Bruce et al. 2019). Approximately 20,000 plant species are used in third world countries (Thampi et al. 2019). Despite the wide use

DOI: 10.1201/9781003173991-2

of medicinal plants, only a small fraction of the estimated plant species worldwide has been investigated phytochemically, and few components have undergone biochemical, biological or pharmacological examinations to validate their activity. The World Health Organization has shown great interest in documenting and recording the use of medicinal plants by peoples around the world, as they are available, low-cost and safe resources if used properly (Irfat et al. 2020; Majid et al. 2021).

In this context, pharmacognosy studies play a fundamental role. They contribute to the standardization of medicinal plants and their final products, through quality assurance practices applied from cultivation (Butt et al. 2018; Ghosh 2018; Nanjan 2018). Advanced analytical techniques serve as rapid and specific tools in herbal medicine research, establishing quality standards and specifications for therapeutic efficacy, shelf-life and safety of herbal products (Bhusnure et al. 2019).

IMPORTANCE OF PHARMACOGNOSY STUDIES

Pharmacognosy is the pharmaceutical science that deals with the study of drugs and medicinal substances of natural origin: vegetable, microbial (fungi, bacteria) and animal. The science explores natural sources of raw materials of pharmaceutical interest; both substances with therapeutic properties and toxic substances, excipients and other useful substances, although their use is basically technological and non-therapeutic. In general, pharmacognosy deals with the botanical, chemical, biological and economic aspects of drugs, intended for the preparation of medicines (Perveen and Mohammad 2019; Cahlíková et al. 2020).

The importance of pharmacognosy lies in the fact that this science allows the development of synthetic biosimilars, which makes it possible to carry out modifications, such as increases in their bioavailability, altered pharmacokinetics and improved efficacy. These modifications can transform an inactive plant into a powerful drug, as has been observed in certain anticancer drugs. Therefore, pharmacognosy studies can provide excellent models for producing new drugs and allow researchers to detect and evaluate the biological properties of plants and determine their effects on living systems (Liji 2021).

In addition, pharmacognosy applies botanical knowledge to classify, name plants and understand their genetic pattern and cultivation. In the field of chemistry, it makes it possible to isolate, identify and quantitatively evaluate bioactive compounds in plant sources. A very important aspect is that pharmacognosy studies ensure the correct identification and purity of the medicinal plant under study, as well as accurate tests of its efficacy and safety (Liji 2021).

Pharmacognosy is a rapidly developing science that deals with the development and use of analytical methods for the quality control of natural products. These include the following:

a) use of traditional remedies by native cultures; the microscopic evaluation and verification of species of medicinal or economically important natural products;
b) use of natural products for specific agricultural purposes, such as natural pesticides or antiseptics; the analysis of the functional and safety properties

of the compounds found in novel foods or food ingredients and consumer products;

c) cosmetic application of natural compounds or extracts and the study and manipulation of genetic biosynthetic pathways in order to improve the production of natural compounds or produce new compounds (Taviad and Vekariya 2018).

In a contemporary context, the use of pharmacognosy has increasingly adopted approaches based on traditional medicine to increase outcomes and address safety concerns. Thus, clinical, analytical and industrial pharmacognosy have established themselves as the specialized and professional branches of pharmacognosy to cope with contemporary advances in this area. Furthermore, molecular, genomic and metabolomic pharmacognosy have been considered promising approaches to pharmacognosy research to accommodate future demands in molecular biology, biotechnology and analytical chemistry of natural drugs and medicinal plants. Thus, interdisciplinary collaborative research programs are essential for the integrated development of pharmacognosy (Taviad and Vekariya 2018).

STANDARDIZATION IN THE PHARMACOGNOSY STUDIES OF PLANT DRUGS

Standardization is a tool in the quality control process that refers to the set of information and control necessary to produce material of reasonable consistency. It is the process involved in the selection and handling of raw material, the evaluation of the safety, efficacy and stability of the finished product, documentation of safety and risk, based on experience, and provision of product information to the consumer and product promotion (Bhusnure et al. 2019; Sachan et al. 2016; Bijauliya et al. 2017). In the case of plant drugs, standardization is not so simple, since the content of herbal preparations depends on the conditions of growth, drying, climate, soil quality, extraction, harvest time, and so on (Nanjan 2018). The standardization of plant drugs refers to the accumulation of information and controls that serve to optimize the consistency from one batch to another of an herbal product. Standardization is achieved by reducing the inherent variation of the composition of the natural product through quality assurance practices applied to agricultural and manufacturing processes (Sachan et al. 2016).

Various factors such as bioefficacy and reproducible therapeutic effects influence the standardization of plant drugs. The main factor under consideration is the adulteration of the herbal ingredient that may be present intentionally or unintentionally, such as lack of storage, mixing of one ingredient with another, similar name of the plants or replacement by similar ones (Parnika and Rakesh 2020).

In plant preparations such as extracts, they must be analyzed to determine the proposed biological activity in an experimental animal model. The active ingredient should be standardized and tested on the basis of the active ingredient or main compound together with the complete drug profile (fingerprints). The next necessary step is the stabilization of the bioactive extract with a minimum shelf life of more than one year. The stabilized bioactive extract should undergo limited regulatory or safety studies in experimental animals. Safe and stable plant extract can be marketed in a suitable formulation if its therapeutic use is well documented in indigenous systems of medicine, as also considered by the World Health Organization (Irfat et al. 2020). Pharmacognosy standardization is an instrument of singular importance to adequately satisfy consumer preferences, as well as to improve the functioning of markets. It contributes to the establishment of a consistent biological activity, a consistent chemical profile, or simply a quality assurance program for the production and manufacture of herbal medicines. It makes it possible to prescribe a set of standards or inherent characteristics, constant parameters, defined qualitative and quantitative values that carry a guarantee of quality, efficacy, safety, stability and reproducibility (Parnika and Rakesh 2020).

MAIN PHARMACOGNOSY INDICATORS IN THE QUALITY CONTROL OF DRUGS AND THEIR EXTRACTS

General test methods for evaluating official drugs are described in various Pharmacopoeias. The World Health Organization has recommended the preparation of monographs and the implementation of quality standards or specifications, based on the experience of each country, for those drugs not included in the Pharmacopoeias and that are widely used in Traditional Medicine. The reliability of the result obtained in the analysis of a sample depends on the selection that has been made of it. Due to the specific characteristics of the drugs, in particular, their lack of homogeneity, special handling procedures related to the taking of samples are required, which can be carried out randomly depending on the total amount of sample available, from which an "average sample" representative of the batch to be analyzed is obtained, with which the quality tests are carried out (Miranda and Cuellar 2012).

Pharmacognostic studies involve several different evaluation techniques that are widely used to determine the quality of natural products. These evaluations focus on botanical characterization; macroscopic or organoleptic evaluation, physicochemical tests; phytochemical evaluation and biological tests (Shailesh et al. 2015; Kumari and Kotecha 2016; Bijauliya et al. 2017; Bhusnure et al. 2019; Oppong et al. 2020; Parnika and Rakesh 2020).

Each monograph contains detailed botanical, macroscopic and microscopic descriptions of the physical characteristics of each plant that can be used to ensure both identity and purity. Each description is accompanied by detailed illustrations and photographic images that provide visual documentation of precisely identified material. According to the WHO, the macroscopic and microscopic account of a plant medicine is the first step to determine the identity and the degree of purity of said material and must be achieved before carrying out any test (Majid et al. 2021).

Botanical Characterization

Botanical characterization is important to identify plant species. It must be done by the scientific name and not by the popular one. The scientific name of plants according to binary nomenclature is made up of the generic name (Latin noun, written with a capital letter) and the name of the species (Latin adjective, written with a lower case) followed by the abbreviated name of the author or discoverer of the species. The family to which the species belongs must also be given and the part of the plant to be used (inflorescences, leaves, roots) must be specified. These aspects are very important since there may be adulterations of the vegetable raw material and if the part of the plant does not correspond to the prescribed, the active principles may not be in the right proportions and may even be devoid of them (Miranda and Cuellar 2012).

On the other hand, genetic analysis has proven to be an important tool in the standardization of medicinal plants. The genotypic characterization of plant species is important since most plants may show considerable variation in morphology, although they belong to the same genus and species. DNA analysis is useful for the identification of cells, individual plants and species, and could help distinguish genuine from adulterated drugs (Kumari and Kotecha 2016).

Macroscopic or Organoleptic Evaluation

Macroscopic evaluation is an effective tool to determine the identity of the species. An evaluation of the plant material is performed, either with the naked eye, with a hand lens or magnifying glass, or with a stereomicroscope. Similar plant species can share similar morphological characteristics and therefore adequate training is needed to acquire skills on the macroscopic identification of plant material (Upton 2009).

Organoleptic evaluation means the study of drugs using organs of senses. It refers to the methods of analysis like colour, odour, taste, size, shape and special features, such as touch, texture and so on (Selvam 2015).

In extracts or other herbal preparations, the smell, color, transparency, presence of particles in suspension or precipitates (if liquid), formation of layers and so on are determined (Miranda and Cuellar 2012).

Microscopic Evaluation

Microscopic evaluation is of interest as it is the most reliable and convenient technique to check the quality of a drug. With this review some essential parameters of plants are described to confirm standard varieties of herbs (Patel et al. 2017).

Advancement in the microscopy technique could help to achieve the goal of standardization of herbal products. Moreover, the majority of regulatory procedures and pharmacopoeias propose macroscopic and microscopic evaluation for herbal standardization. To ensure the quality of plant material, microscopic description is the first step towards establishing its identity and purity (Patel et al. 2017).

This method allows more detailed examination of a drug and it can be used to identify the organized drugs by their known histological characters. It is mostly used for qualitative evaluation of organized crude drugs in entire and powdered forms. Every plant possesses a characteristic tissue feature. A microscope can be used to confirm the structural details of the drugs from plant origin. For the effective results, various reagents or stains can be used to distinguish cellular structure.

Micromorphological characteristics of the leaf are more taxonomically useful than gross or external morphology. Various leaf characteristics have been applied in systematic studies for different taxonomic groups and have also been used to facilitate accurate authentication and quality control of medicinal plants (Song et al. 2020).

Determination of leaf constants include stomatal number, stomatal index (percentage of stomata in relation to the total number of epidermal cells, each stoma being counted as one cell), vein islet (number of vein islet per sq. mm of the leaf surface midway between the midrib and the margin; a constant for a given species of plant), veinlet termination number (number of veinlet termination per sq. mm of the leaf surface midway between midrib and margin) and palisade ratios (the average number of palisade cells beneath each epidermal cell) (Patel et al. 2017).

Stomata are minute pores which occur in the epidermis of the plants. Each stoma remains surrounded by two kidneys or bean shaped epidermal cells – the guard cells. Stomata may occur on any part of a plant except the roots. The epidermal cells bordering the guard cells are called accessory cells or subsidiary cells. Stomata is one of the best standards to identify the plant microscopically, and may vary according to the plant species (Patel et al. 2017; Patel et al. 2020).

Cell structures to observe in a drug test can include cell walls, parenchymal tissues, epidermis, epidermal trichomes, endodermis, cholenchyme, sclereids, fibers, xylem, phloem, secretory tissues, starch grains, and more (Patel et al. 2017).

Trichomes are outer growth of epidermal cells mostly found on leaves and stem, sometimes on fruits and flowers. Trichomes are highly varied amongst the plants and their diversity as well as distribution have been used as important tools in delimiting plant species. Foliar trichome character has been used to resolve taxonomic conflicts and have played an important role in plant taxonomy. Trichomes are an important tool to identify herbs in powders as well to check for adulteration. They are of scientific interest due to their functional attributes and economic importance in the secretion of phytochemicals (Patel et al. 2017; Patel et al. 2020).

Xylem is one of the essential tissues of the vascular system of plants. It transports water upwards. Plants have specific type of xylem vessels according to lignin present on its walls. One can easily check the purity of drug powder in microscopy by patterns of xylem vessels in comparison to standard literature (Patel et al. 2017).

Starch analysis or starch grain analysis is a technique that is useful in archaeological research to determine plant taxa. Plant starch grain analysis is a diagnostic feature of multiple applications according to the peculiarities and origins of the plant material. The size, shape and structure of grains from plant species varies little, which can lead to identification (Patel et al. 2017).

It is also important to determine the ergastic cellular content that represents food storage products or byproducts of metabolism and includes carbohydrates, proteins, fixed oils and fats, alkaloids and purines, glycosides, volatile oils, gums and mucilage, resins, tannins, calcium oxalate and silica.

Microscopic evaluation of plant material is crucial for the detection of parent materials. Anatomical attributes are used as criteria to unravel species, genera and even families. In addition, anatomy gives diagnostic characteristics of plant material for quality control and standardization (Majid et al. 2021).

Physicochemical Evaluation

Physicochemical evaluation informs about the identity of the material and constitutes an initial test for detection of impurities. Among the parameters to be measured are (Majid et al. 2021):

Determination of foreign matter: involves the removal of matter other than the plant of origin to obtain the drug in pure form. Examples include other parts of the plant or other plant, dust, sand, stone, harmful and poisonous foreign matter and chemical residues, insects, and other animal contamination, including animal excreta (Kshirsagar et al. 2017).

Determination of ash: The ash remaining following ignition of medicinal plant materials is determined by 3 different methods which measure total ash, acid-insoluble ash and water-soluble ash. The total ash method is designed to measure the total amount of material remaining after ignition. This includes both "physiological ash", which is derived from the plant tissue itself, and "non- physiological" ash, which is the residue of the extraneous matter adhering to the plant surface. Acid-insoluble ash is the residue obtained after boiling the total ash with dilute hydrochloric acid, and igniting the remaining insoluble matter. This measures the amount of silica present, especially as sand and siliceous earth. Water soluble ash is the difference in weight between the total ash and the residue after treatment of the total ash with water (WHO 2011; Bijauliya et al. 2017).

Moisture content: helps reduce errors in estimating the actual weight of plant material. Low humidity suggests better stability against product degradation. This determination is carried out on both raw drugs and dry extracts (WHO 2011; Miranda and Cuellar 2012).

Extractable matter: this method determines number of active constituents extracted with solvents from a given amount of medicinal plant material (Bijauliya et al. 2017; Oppong et al. 2020).

Crude fiber: helps determine the component of the woody material and is a criterion for judging purity.

Fluorescence analysis: Certain drugs fluoresce when the cut surface or the powder is exposed to ultraviolet radiation, and it is useful in the identification of those drugs.

For the quality of extracts also are parameters to be measured:

Density or specific weight (for liquids): is the ratio between the mass of a volume of the substance to be tested at 25°C and the mass of an equal volume of water at the same temperature (Miranda and Cuellar 2000).

Refraction index (for liquids): is a constant characteristic of each substance, which represents the relationship between the sine of the angle of incidence of light and the sine of the angle of refraction when light passes obliquely through the medium. In practice refractometers are used (Miranda and Cuellar 2000).

pH determination: measures the degree of acidity or alkalinity of the solutions. In practice, a pH measuring instrument is used, either digital or analog (Miranda and Cuellar 2000).

Total solids: determines the variation of the mass, due to the loss or elimination of volatile substances due to the action of heat, through a process of evaporation of the test portion and drying of the residue in an oven, until a constant mass is achieved (Miranda and Cuellar 2000). The percentage obtained provides information on the amount of non-volatile solids (including secondary metabolites responsible for therapeutic action) present in an extract, in addition to serving as a basis for pharmacologists to adjust the dose when it is evaluated in pharmacological and toxicological tests.

Capillary analysis: is an ingenious and interesting method for the characterization of extracts and dyes. This method is based on the phenomena of absorption and distribution of substances in coloring materials through the capillary spaces of the inert material that constitutes the filter paper. This analysis provides information on possible alkalinity changes when the sample is exposed to ammonia vapors and the presence of compounds with fluorescent characteristics when analyzed under ultraviolet light (Miranda and Cuellar 2000).

Alcohol content: consists of obtaining a hydroalcoholic solution through the sample distillation process, where it must be guaranteed that all the alcohol present is drawn into the distillate. With the subsequent determination of the specific gravity of said distillate, the percentage of alcohol in the sample is obtained. The percentage of ethanol determined must be close to the percentage of ethanol used to obtain the extract or tincture (Miranda and Cuellar 2000).

Other physicochemical variables include *viscosity*, *melting point*, *optical rotation*, *solubility* and more.

Physical standards are to be determined for the drugs, wherever possible. These are rarely constant for crude drugs, but may help in evaluation.

Chemical Evaluation

The chemical evaluation includes qualitative and quantitative tests, chemical assays and instrumental analysis. The isolation, purification and identification of active constituents are chemical methods of evaluation. Qualitative chemical tests include identification tests for various phytoconstituents (Bijauliya et al. 2017).

Fingerprint profiles are used as a guide for the phytochemical profile of the drug to ensure quality, while quantification of marker compounds would serve as an additional parameter to assess the quality of the sample. Phytochemical standardization encompasses all possible information generated regarding the chemical components present in an herbal product (Bijauliya et al. 2017).

Markers are chemically defined constituents of an herbal drug which are of interest for quality control purposes independent of whether they have any therapeutic activity or not. Markers may serve to calculate the amount of active component of herbal drug or preparation in the finished product (Pintoa et al. 2015).

Marker compounds are pure, single isolated compounds, secondary metabolites mostly with terpenes, steroid, alkaloid, flavonoid, aromatic, hetero aromatic frameworks and glycosides having alcoholic, carbonyl, olefinic, acid, ester and amide functionalities highly useful for single/crude drugs: may or may not survive in multiherbal. For quantitative studies, use of specific markers that can be easily analyzed to distinguish between varieties remains a preferred option (Pintoa et al. 2015).

The quantity of a chemical marker can be an indicator of the quality of an herbal medicine. The study of chemical markers is applicable to many research areas, including authentication of genuine species, search for new resources or substitutes of raw materials, optimization of extraction and purification methods, structure elucidation and purity determination. Systematic investigations using chemical markers may lead to discoveries and development of new drugs (Li et al. 2008; Rasheed et al. 2012).

Control tests on the finished product will allow the quantitative and qualitative determination of the composition of the active ingredients and will give a specification using markers if the constituents with therapeutic activity are unknown. In the case of compounds of known therapeutic activity, these constituents will have to be specified and quantitatively determined. If an herbal medicine contains various herbal drugs or preparations of various herbal drugs and it is not possible to perform a quantitative determination of each active ingredient, the determination will be carried out for the mixture.

Critical to compliance with any monograph standard is the need for adequate analytical methods to determine identity, quality and relative potency. These methods include: ultraviolet spectroscopy, infrared spectroscopy, thin layer chromatography, high performance thin layer chromatography, high performance liquid chromatography, gas chromatography, liquid chromatography-mass spectrometry, gas chromatography-mass spectrometry and nuclear magnetic resonance, among others. The results of these sophisticated techniques provide a chemical fingerprint as to the nature of the chemicals or impurities present in the plant, extract, or finished product (Balekundri and Mannu 2020).

Biological Evaluation

When the estimation of potency of a crude drug or its preparation is done by means of its effect on living organisms like bacteria, fungal growth or animal tissue or entire animal, it is known as bioassay. This method is generally called for when standardization is not adequately done by chemical or physical means and also for conformity of therapeutic activity of raw material and finished product. In other words, bioassay is the measure of sample being tested capable of producing biological effect as that of the standard preparation.

Pharmacological screening constitutes one of the initial stages in research on medicinal plants and seeks to discover those that have pharmacological activity through carefully performed techniques, so that it is safe and reproducible; however, the techniques and procedures should not be overly elaborate and expensive. Procedures should be programmed in such a way that raw material, such as plant extracts or extract fractions, can be used. Validated pharmacological models should be used, in addition to positive controls, the use of a wide dose range and the application of good laboratory practices and ethical principles for working with biological reagents and laboratory animals.

On the other hand, *toxicological studies* are important to decide whether a new drug should be modified for clinical use or not, to provide researchers with information on the doses from which toxic effects begin to appear. Depending on the period of contact with animals to a drug, toxicological studies may be of 3 types: acute, subacute and chronic. Toxicity depends not only on the dose of the substance but also on the toxic properties of the substance (Bhusnure et al. 2019).

Other important tests that are carried out on drugs from a biological point of view are:

Determination of bitterness value: allows to estimate the value of the aqueous extract of the plant as a tonic for the appetite due to the presence of bitter principles, which stimulate the secretions of the gastrointestinal tract and in particular of gastric juice (Bijauliya et al. 2017).

Foaming Index: This method involves persistent foam by the aqueous decoction of the herbal material and their extract, which helps in analyzing saponin content in the medication (Bijauliya et al. 2017; Parnika and Rakesh 2020).

Swelling Index: This index is useful in the evaluation of crude drugs containing mucilage, gums, pectin and hemicellulose. The swelling index is the volume in ml taken up by the swelling of 1 gm of plant material under specified conditions. Its determination is based on the addition of water or a swelling agent as specified in the test procedure for each individual plant material (Bijauliya et al. 2017; Parnika and Rakesh 2020).

Haemolytic activity: Haemolysis is caused by the inherent property of saponins. The haemolytic activity of plant materials, or a preparation containing saponins, is determined by comparison with that of a reference material, saponin, which has a haemolytic activity of 1,000 units per g. A suspension of erythrocytes is mixed with equal volumes of a serial haemolysis and is determined after allowing the mixtures to stand for a given period of time. A similar test is carried out simultaneously with saponin (Bijauliya et al. 2017; Parnika and Rakesh 2020).

Determination of pesticide residues: Herbal drugs are prone to contain pesticide residue, which accumulate from agricultural practices such as spraying as well as behavior of soil during cultivation and addition of fumigants during storage. Most pesticides contain organically bound chlorine or phosphorus (Bijauliya et al. 2017; Parnika and Rakesh 2020).

Determination of microorganisms and aflatoxins: Microbial parameters, such as total viable content of pathogenic bacteria like enterobacteria (gram negative bacteria including *Escherichia coli*, *Enterobacter*, *Klebsiella*, *Pseudomonas aeruginosa*, *Salmonella*, *Shigella*), and other gram-positive bacteria are determined. The occurrence of fungi should be monitored, to prevent the species mutagen (*Aspergillus*, *Fusarium*), which produce aflatoxins. Absorption of minute amount of aflatoxin is hazardous to humans.

Limits given in official books can be utilized as a quantitative or semi quantitative tool to control the amount of impurities coming from different steps of preparation, storage and preservation (Bijauliya et al. 2017; Parnika and Rakesh 2020).

LABELING, STORAGE AND STABILITY IN THE PHARMACOGNOSTIC STUDY OF HERBAL PRODUCTS

Labeling of Herbal Products

An important aspect in the pharmacognostic standardization of herbal products is the labeling. The quality of consumer information about the product is as important as the finished herbal product. Warnings on the packet or label will help to reduce the risk of inappropriate uses and adverse reactions. The primary source of information on herbal products is the product label. Studies of herbal products have shown significant differences between what is listed on the label and what is in the container (Kunle et al. 2012; Kr Sachan et al. 2016).

To avoided confusion medicinal plant drug material should be well labelled on the packaging with clear indication about botanical name, biological source, plant structure, its assortment site, dates, and names of the grower, collector and processors, as well its quantity. Additional labelling about quality approval should be in accordance with various required national and/or regional standards, giving extra authentic advantage to medicinal drug plants (Porwal et al. 2020).

Labeling on herbal formulations must contain: the name of the product; the declaration of active ingredients; the identification of the manufacturer and, where appropriate, the distributor; the instructions for its conservation; the expiration date; the batch number; the dose and route of administration; the precautionary legends, including risk of use in pregnancy; warning legends; and so on.

Storage of Herbal Products

The storage and conservation of medicinal plants have the purpose of slowing the deterioration of its quality and maintaining the qualitative and quantitative aspects after drying, through the ideal conditions of temperature and relative humidity, to prevent the attack of microorganisms, fungi and insects during the period of storage. During storage, metabolic activity must be reduced, making medicinal plants less susceptible to deterioration. This can be achieved by reducing the water content of the product to safe levels, cooling, or use of modified atmosphere in the system in which the medicinal plants are stored (WHO 2011; Cristiane et al. 2018).

Storage warehouses are preferably built of non-combustible material such as: steel, bricks and so on, they must be ventilated and have controlled temperature and humidity. It is also important to guarantee the absence of rodents. Except in some cases, prolonged storage, although preferably unavoidable, is detrimental. Certain drugs deteriorate even when carefully stored. Drugs stored in the usual containers (sacks, bundles, wooden crates, cardboard boxes, and paper bags), absorb approximately 10 to 12% more moisture. During drug storage, the combined effects of the drug's own humidity, with which the temperature exerts on the environment and subsequent water condensation when lowering the temperature, must be taken into account (Miranda and Cuellar 2012).

Some drugs should never take moisture from the air, as they lose a considerable part of their activity. They should be put in airtight containers with a dehydrating agent. For large quantities, a drawer with quicklime at the bottom and a grid can be used to separate the drug from the lime; it must be renewed as it becomes wet. Air-dried drugs are always exposed to attack by insects and other pests, so they must be frequently examined during storage (Miranda and Cuellar 2012).

STABILITY

Stability testing is an important component of herbal drugs and products development processes. Stability studies on herbal drugs reported in literature involve quantitative monitoring of specific constituent(s) as active and/or analytical marker(s). However, any change in content(s) of a specific or a group of specific marker(s) in an herbal product during its stability studies may not be extrapolated to similar changes in its therapeutic effectiveness. It is because an herb in its entirety is regarded as the active ingredient. Constituents belonging to different chemical classes in an herbal product may undergo varied intramolecular or intermolecular reactions under the influence of heat, humidity and/or light experienced during its manufacture, transportation and storage (Bansal et al. 2016).

The possible interactions between different groups of constituents are liable to produce products that consequently may render the product more or less active and/or toxic. Such interactions are liable to change the chemical profile of an herbal drug during its storage, which may, consequently, affect or alter its therapeutic profile. Therefore, it is necessary to show, through chromatographic fingerprints generated by sophisticated analytical methods, that the overall chemical composition of a drug remains unchanged during its shelf life. In addition, herbal drugs are commonly formulated and dispensed as tablets, capsules, topical products and oral solutions. Various parameters specific to these formulations such as disintegration, dissolution, hardness, brittleness, pH, viscosity, clarity, suspendability, homogeneity, extractables and microbial contamination do change with time during storage. Hence, these parameters are also required to be evaluated during their stability testing as per the pharmacopeial protocols. However, it is rather unnecessary to undertake all the listed tests for each type of product; the tests undertaken should be able to justify the overall stability studies. Therefore, stability studies of an herbal product constitute an indispensable component of its development process (Bansal et al. 2016).

Stability testing helps in establishing the storage conditions of a drug product throughout its shelf life. The influence of various storage conditions is much more complicated for an herbal product than that for a synthetic drug product. The major challenges that make stability testing of an herbal drug/product include its chemical complexity, variability in biochemical composition of raw material, selection of marker(s) for its stability testing and influences of enzymes present (Bansal et al. 2016).

An herbal drug/product is a very complex mixture of chemicals, and its therapeutic actions are usually a function of additive or synergistic actions of chemically diverse phytoconstituents. It implies that any change in content of a specific marker or a set of specific markers during stability testing of an herbal drug/product is not likely to transcend to similar change in its therapeutic effectiveness (Bansal et al. 2016). It is recommended that, in case of an herbal medicinal product containing a natural product or an herbal drug preparation with constituents of known therapeutic activity, the variation in component during the proposed shelf-life should not exceed ± 5% of the initial assay value, unless justified to widen the range up to ± 10% or even higher (Ghate et al. 2019). Therefore, selection of constituents as markers is the most challenging task in rational stability testing protocol for assessment of shelf life of an herbal drug/product (Bansal et al. 2016).

In order to establish safe and therapeutic efficacy of an herbal drug product, an evaluation of influence of storage conditions is required not only on physicochemical stability but also on biological activity, toxicity and microbial contamination. Hence, there is a need for mandatory evaluation of physical and chemical stability along with major biological activities of an herbal drug/product during systematic stability studies (Bansal et al. 2016).

CONCLUSION

Pharmacognostic studies applied to the quality control of plant drugs contemplate a series of laboratory experiments that reveal and assemble a set of constant parameters, and define qualitative and quantitative values, or specific characteristics, that become an important measure to guarantee quality, purity and authenticity of herbal products. In this sense, standardization plays an essential role since it allows obtaining effective and safe products of good quality.

Evaluating plant drugs involves identifying them and determining their quality and purity, which translates to their intrinsic value in the amount of active principles present.

At present, phytochemical research requires the determination of chemical composition and its relationship with biological activity. In this sense, current trends as a result of modern isolation and pharmacological experimentation procedures are aimed at identifying the metabolites responsible for therapeutic utility and, on that basis, being able to achieve pharmaceutical preparations that meet the safety and efficacy standards required by WHO or international monographs in a general sense. The chemical composition determined depends on the quality of the drug used, but in turn this would lead to a pharmacognostic investigation that would influence the variability of the raw material in relation to intrinsic and extrinsic factors.

It is recommended that various government agencies follow a more universal approach to the quality of herbal products by adopting the WHO guidelines and also producing monographs using the various quality parameters described above. This will strengthen the regulatory process and minimize quality non-compliance.

REFERENCES

Balekundri, A. and Mannu, V. 2020. Quality control of the traditional herbs and herbal products: a review. *Future Journal of Pharmaceutical Sciences* 6:67. https://doi.org/10.1186/s43094-020-00091-5

Bansal, G., Suthar, N., Kaur, J. and Jain, A. 2016. Stability testing of herbal drugs: challenges, regulatory compliance and perspectives. *Phytotherapy Research* 30(7):1046–1058. https://doi.org/10.1002/ptr.5618

Bhusnure, O. G., Suryawanshi, S., Vijayendra, S. S. M, Gholve, S. B, Girm, P. S. and Birajdar, M. J. 2019. Standardization and quality evaluation of herbal drugs. *Journal of Drug Delivery and Therapeutics* 9(3-s):1058–1063. https://doi.org/10.22270/jddt.v9i3-s.2941

Bijauliya, R. K., Alok, S., Chanchal, D. K. and Kumar, M. 2017. A comprehensive review on standardization of herbal drugs. *International Journal of Pharmaceutical Sciences and Research (IJPSR)* 8(9):3663–3677. https://doi.org/10.13040/IJPSR.0975-8232

Bruce, S. O., Onyegbule, F. A. and Ezugwu, C. O. 2019. Pharmacognostic, physicochemical and phytochemical evaluation of the leaves of Fadogia cienkowski Schweinf (Rubiaceae). *Journal of Pharmacognosy and Phytotherapy* 11(3):52–60. https://doi.org/10.5897/JPP2019.0552

Butt, J., Ishtiaq, S., Ijaz, B., Mir, Z. A., Arshad, S. and Awais, S. 2018. Standardization of herbal formulations at molecular level. *Journal of Pharmacognosy & Natural Products* 4(1):1000150. https://doi.org/10.4172/2472-0992.1000150

Cahlíková, L., Šafratová, M., Hošt'álková, A., Chlebek, J., Hulcová, D., Breiterová, K., et al. 2020. Pharmacognosy and its role in the system of profile disciplines in pharmacy. *Natural Product Communications* 15(9):1–7. https://doi.org/10.1177/1934 578X 20945450

Cristiane, F. L., Evandro, C. M. and Sérgio, M. L. D. 2018. Influence of storage conditions on quality attributes of medicinal plants. *Biomedical Journal of Scientific & Technical Research* 4(4):4093–4095. BJSTR. MS.ID.001097. https://doi.org/10.26717/BJSTR.2018.04.001097

Ghate, V. U., Gajendragadkar, M. P. and Jadhav, A. B. 2019. Importance of standardization of herbal drug in homoeopathy. *International Journal of Research* VIII (II):870–879.

Ghosh, D. 2018. Quality issues of herbal medicines: internal and external factors. *Journal of Integrative and Complementary Medicine* 11(1):67–69. https://doi.org/10.15406/ijcam.2018.11.00350.

Irfat, A., Mudasir, M., Basharat, B., Nighat, A. and Tawseef, A. H. 2020. Present status, standardization and safety issues with herbal drugs. *International Journal of Pharmaceutical Sciences and Research* 1(3):98–101. https://doi.org/10.33974/ijrpst.v1i3.169

Jamshidi, K. F., Lorigooini, Z. and Amini, K. H. 2018. Medicinal plants: past history and future perspective. *Journal of Herbmed Pharmacology* 7(1):1–7.

Kr Sachan, A., Vishnoi, G. and Kumar, R. 2016. Need of standardization of herbal medicines in modern era. *International Journal of Phytomedicine* 8:300–307.

Kshirsagar, D. P., Gaikwad, S. S., Karale, P. R., Vipul, P. and Rasika, B. 2017. Current quality control methods for standardization of herbal drugs. *International Journal of Pharmaceutics and Drug Analysis* 5(3):82–95.

Kumari, R. and Kotecha, M. 2016. A review on the Standardization of herbal medicines. *International Journal of Pharma Sciences and Research* 7(2):97–106.

Kunle, O. F., Egharevba, H. O. and Ahmadu, P. O. 2012. Standardization of herbal medicines—A review. *International Journal of Biodiversity and Conservation* 4(3):101–112. https://doi.org/10.5897/IJBC11.163

Liji, T. 2021. *Importance of pharmacognosy.* Available at: www.news-medical.net/health/Importance-of-Pharmacognosy.aspx

Li, S., Han, Q., Qiao, C., Song, J., Lung, C., Cheng, C. L., et al. 2008. Chemical markers for the quality control of herbal medicines: an overview. *Chinese Medicine* 3:7.

Majid, N., Nissar, S., Younus, R. W., Nawchoo, I. A. and Ali, B. Z. 2021. Pharmacognostic standardization of Aralia cachemirica: a comparative study. *Journal of Pharmaceutical Sciences* 7:33. https://doi.org/10.1186/s43094-021-00181-y

Miranda, M. M. and Cuellar A. C. 2000. Laboratory practice manual. *Pharmacognosy and natural products*. Havana: Editorial Felix Varela, 25–49, 74–79.

Miranda, M. M. and Cuellar A. C. 2012. *Pharmacognosy and chemistry of natural products*. Havana: Editorial Felix Varela. Second edition, 135–145.

Nanjan, M. J. 2018. Standardization of herbal drugs: a critique. *Journal of Natural & Ayurvedic Medicine* 2(1):000121.

Oppong, B. E., Dodoo, B. K., Kitcher, C., Gordon, A., Frimpong, M. S. and Schwinger, G. 2020. Pharmacognostic characteristics and mutagenic studies of *Alstonia boonei* De Wild. *Research Journal of Pharmacognosy* 7(1):7–15. https://doi.org/10.22127/rjp.2019.204629.1526

Parnika, C. N. and Rakesh, K. S. 2020. Review article: standardization and evaluation of herbal drugs is need of the hour in present era. *Plant Archives* 20(2):7883–7889.

Patel, M. T., Purohit, K. and Patel, M. R. 2017. Microscopic evaluation: an essential tool for authentification of crude drugs. *World Journal of Pharmaceutical Research* 6(17):334–343. https://doi.org/10.20959/wjpr201717-10374

Patel, S. R., Joshi, A. G., Pathak, A. R. and Raole, V. M. 2020. Micromorphology studies of three important medicinal plants of Asclepiadaceae family. *Notulae Scientia Biologicae* 12(1):22–29. https://doi.org/10.15835/nsb12110528

Perveen, S. and Mohammad, Al-Taweel, A. 2019. Introductory chapter: pharmacognosy. in: pharmacognosy—medicinal plants. *IntechOpen*:1–8. https://doi.org/10.5772/intechopen.86019

Pintoa, R. M. C., Lemesa, B. M., Zielinskib, A. A. F., Kleina, T., de Paulac, F., Kistd, A., et al. 2015. Detection and quantification of phytochemical markers of */lex paraguariensis* by liquid chromatography. *Química Nova* 38(9):1219–1225. https://doi.org/10.5935/0100-4042.20150117

Porwal, O., Kumar, S. S., Kumar, P. D., Gupta, S., Tripathi, R. and Katekhaye, S. 2020. *Cultivation, collection and processing of medicinal plants*. Bioactive Phytochemicals: Drug Discovery to Product Development, 1–16.

Rasheed, N. M. A., Nagaiah, K., Goud, P. R. and Sharma, V. U. M. 2012. Chemical marker compounds and their essential role in quality control of herbal medicines. *Annals of Phytomedicine* 1(1):1–8.

Selvam, A. B. D. 2015. Standardization of organoleptic terminology with reference to description of vegetable crude drugs. *International Journal of Pharmacy and Technology* 7(2):3282–3289.

Shailesh, L. P., Suryawanshi, A. B., Gaikwad, M. S., Pedewad, S. R. and Potulwar, A. P. 2015. Standardization of herbal drugs: an overview. *The Pharma Innovation Journal* 4(9):100–104.

Song, J. H., Yang, S. and Choi, G. 2020. Taxonomic implications of leaf micromorphology using microscopic analysis: a tool for identification and authentication of korean piperales. *Plants* 9:566. https://doi.org/10.3390/plants9050566

Taviad, K. and Vekariya, S. 2018. The scope of pharmacognosy today & tomorrow. *International Journal of Pharmacognosy & Chinese Medicine* 2(1):000127.

Thampi, R., Mercykutty, M. J. and Menon, J. S. 2019. Traditional knowledge on use of medicinal plants grown in homesteads as home remedies. *Journal of Medicinal Plants Studie*s 7(2):1–4.

Upton, R. 2009. Authentication and quality assessment of botanicals and botanical products used in clinical research. *Evaluation of herbal medicinal products perspectives on quality, safety and efficacy*. Edited by Pulok K Mukherjee and Peter J Houghton. Pharmaceutical Press. Trowbridge: Printed in Great Britain by Cromwell Press Group, 383–391.

World Health Organization (WHO). 2011. *Quality control methods for herbal materials*. ISBN 978 92 4 150073 9.

2 Application of DNA Barcodes to Medicinal Plants

Efrén Santos, Ricardo Pacheco and Liliana Villao

CONTENTS

INTRODUCTION

Since the development and implementation of nucleic acid sequencing, genetic material could be analyzed for the identification and biodiversity studies of species. Therefore, DNA sequence analysis could be exploited to complement species identification, as sequences could be viewed as "barcodes", which are embedded in the cell (Herbert et al. 2003). Morphological analysis for species identification could be difficult for some organisms, due to different limitations including: i) phenotypic; ii) morphological cryptic taxa could be common in some groups; and, iii) the use of keys could demand a high level of expertise (Herbert et al. 2003). In many animal groups, the use of mitochondrial genes, including CO1, is routinely used for identification and biodiversity analysis (Herbert et al. 2003). However, in plants, low substitution rates in mitochondrial DNA have led researchers to look for other loci, including plastid DNA (Kress et al. 2005; Chase et al. 2007; Ford et al. 2009). For appropriate loci selection, different parameters should be tested including universality, sequence quality and coverage and discrimination (CBOL Plant Working Group 2009). In this chapter, the application of DNA barcodes is discussed for proper identification and diversity analysis of medicinal plants, describing a protocol for DNA extraction and PCR amplification of plastid and nuclear DNA sequences.

DOI: 10.1201/9781003173991-3

USE OF DNA SEQUENCES FOR CHARACTERIZATION OF PLANT SPECIES

Many organisms are characterized by the analysis of DNA sequences (Hebert et al. 2003). These studies have improved the accuracy with which species could be identified, allowing scientists to gain a more in-depth understanding of evolutionary history. The ability to identify the genetic diversity of plant populations is also invaluable for management decisions that involve managing or protecting local biodiversity.

Several benefits could be encountered using DNA barcodes, as this methodology has been shown to be more accurate than morphological characteristics for the identification of plant species; but, more importantly, could serve as a complement analysis for taxonomic identification. This is mainly because many species of plants are very similar in appearance but could be very different genetically. DNA barcodes allow scientists to overcome this problem and accurately identify plant species in a manner that is cost-effective.

More accurate identification of plant species also allows researchers to better document the plant diversity in different ecosystems (Ajmal et al. 2014), therefore contributing to our knowledge of evolution and natural selection. Furthermore, DNA barcodes are an important tool for conservation and use of plant biodiversity. In areas that are being deforested, DNA barcodes could be used to help alert management agencies that new species may be present in the area and should therefore be protected. DNA barcodes could also be used to monitor the spread of invasive species of plants, and to determine whether or not certain species are becoming at risk for extinction.

The low cost of DNA barcoding makes it ideal for use in several applications. A single DNA barcode will identify a plant's genus or species, depending on which sequence is used. For instance, the chloroplast matK could be compared as an analogue to CO1, which is common animal barcode (Srivastava et al. 2016). A criterion should also include a minimum sequence length of 500 bp and more than 3 specimens per species (Sarwat and Yamdagni 2016). However, other strategies for DNA barcode analysis include multilocus and tiered approaches (reviewed by Srivastava et al. 2016). However, a single-locus approach could not be universally applied in plants as, for instance, by using matK due to unavailability of universal primers for all taxa (Kress and Erickson 2007; CBOL Plant Working Group 2009). Therefore, the barcode rbcL is indicated as a complement sequence for species discrimination as a two-locus barcode together with matK (CBOL Plant Working Group 2009). The CBOL Plant Working Group (2009) tested different sequences for barcodes in plants including *atpF-atpH* spacer, *matK* gene, *rbcL* gene, *rpoB* gene, *rpoC1* gene, *psbK—psbI* spacer and *trnH—psbA* spacer.

Based on different assessments including recoverability, sequence quality and species discrimination, the authors recommended the 2-locus combination rbcL+matK as a plant barcode. Other barcodes have been tested in plants, including *rpl32-trnL*, *rps16-trnK* and *trnS-trnFM* (Armstrong et al., 2014); *trnH-psbA* spacer, *trnC-trnD* region, *trnC-psbM* region, and the 3′ end of *ndhF* (Richardson et al., 2013). However, for the cases that studies rely on the comparison of sequence availability in the GenBank, the rbcL, matK, ITS1 and ITS2 are more available.

Although CBOL Plant Working Group (2009) suggest the use of rbcl+matK, several studies revealed that for some species the rbcL+matK could not be used for species discrimination, including medicinal plants; while the ITS sequences (ITS1 and/or ITS2) could be suitable for discrimination at species level (Techen et al 2014; Zhang et al. 2016; Bustamante et al. 2019; Sarmiento-Tomalá et al. 2020; Soledispa et al. 2021). However, some studies report drawbacks for using ITS, as sequencing could be difficult in some plant species (Sass et al. 2007; Tripathi et al. 2013; Gonzalez et al. 2009); that could be due to secondary structure formations, multiple copies of paralogs, fungi contamination, and/or variation in sequence lengths, resulting in a decrease of sequence quality (Hollingsworth 2011). On the other hand, several studies indicated that the ITS could be used as a complement to core barcodes due to a better resolution at species level (Li et al. 2011; Vivas et al. 2014; Tripathi et al. 2013; Gonzalez et al. 2009). In a tiered approach; first, a coding region, for instance rbcL, provides resolution at a genus level. Then, a more variable region (coding or noncoding) provides resolution at the species level. An example of a tiered approach includes the rbcl+ITS2 (Purushothaman et al. 2014). Also, the matK is suggested as a first-tier locus as indicated by Xiang et al. (2011).

DNA BARCODE IN MEDICINAL PLANTS

Among human civilizations various tissues from plants have been used to treat illness or diseases. Nowadays, the global market for medicinal plants is increasing rapidly. The need for proper identification of plant species used for consumption is a key issue to be addressed, as globally, substitutes or adulteration is a major problem in the plant medicinal market (Kumari and Kotecha 2016; Srivastava et al. 2016; Yu et al. 2021). Identification of plant species in markets, including medicinal, could be challenging as plant tissues are often dried or processed. Therefore, DNA barcodes could aid in the identification and differentiation of plant species, in combination with morphology and traditional knowledge analysis (Ghorbani et al. 2017). Furthermore, an integrated approach including taxonomy and DNA barcode increases the efficiency for the identification of species (Dasmahapatra and Mallet 2006; Ajmal et al. 2014). In medicinal plants, DNA barcode could be used as a complement analysis for species identification; where different barcodes were reported (reviewed by Techen et al. 2014) including rbcL, matK, ITS1 and ITS2. Furthermore, the ITS has the potential to distinguish medicinal and close relatives (Kim et al. 2016).

Other locus combination could be used instead of the recommended rbcl+matK; for instance, the rbcL + ITS2 is effective for documenting plant diversity (Liu et al. 2015). The ITS barcode in combination with other sequences (rpoC1, psbA-trnH) were determinant for genus identification in a market for medicinal plants in Southern Morocco (Kool et al. 2012). Furthermore, the implementation of DNA barcodes will determine proper labeling due to misidentification and deliberate adulteration of plant species in the herbal/medicinal industry (Kool et al. 2012; Urumarudappa et al. 2016, 2019), the ITS being a suitable barcode for species differentiation (Li et al. 2013; Zhou et al. 2014; Ganie et al. 2015; Intharuksa et al. 2020). On the other hand, combination with other techniques is suggested for proper identification including omics tools such as transcriptomics, proteomics, and metabolomics (Mishra et al. 2016) or NMR spectroscopic (Urumarudappa et al. 2016).

A STANDARDIZED METHODOLOGY FOR DNA BARCODE IN MEDICINAL PLANTS

Sampling and DNA Extraction

Depending on the morphological characteristics of medicinal plants, different tissues could be used for DNA extraction. The preferred tissue is the leaves; however, any other tissue could be used, although DNA extraction should be optimized depending on the tissue. Furthermore, medicinal plant-based products should contain degraded tissues/DNA, and specialized protocols could be used including CTAB, silica- or streptavidin-based.

A modified CTAB protocol is indicated as follows. Briefly, tissues are harvested and could be stored at −80°C upon DNA extraction. Once the tissues are stored at -80°C, the cold chain is maintained upon incubation with extraction buffer to avoid DNA degradation. Approximately, 100–120 mg of tissue is ground, either using a grinder (MM400, Retsch) or mortar with liquid nitrogen. If liquid nitrogen is not available, samples should be ground together with extraction buffer. The ground powder is transferred to a 2 mL microcentrifuge tube containing 800 µL of extraction buffer (CTAB 20 g/L, 1.3 M NaCl, 0.1 M Tris base/HCl, 20 mM Na2-EDTA pH 8.0) and vortex. A volume of 20 µl of Proteinase K (20 mg/ml) is added and mixed by vortex. The tubes are incubated for 1 hour at 65°C (mixing every 10 min) and, then, the tubes stand for 5 minutes at room temperature. A volume of 700 µl of chloroform:isoamyl alcohol (24:1) is added, and the tubes are mixed by inverting several times. Then, samples are centrifuged for 10 minutes at 14.000 rpm at 4°C (for all the centrifugation steps, if a refrigerated centrifuge is not available, centrifugation could be performed at room temperature). Carefully, the supernatant is transferred to a new tube. RNase is added (final concentration of 200 µg/mL) and incubated for 15 minutes at 37°C. One volume of chloroform:isoamyl alcohol (24:1) is added. Samples are centrifuged for 10 minutes at 14.000 rpm at 4°C. The supernatant is transferred to a new tube. One volume of cold isopropanol is added and stored at -20°C (for better results keep at -20°C overnight). Centrifugation goes for another 20 minutes at 14.000 rpm at 4°C. The supernatant is removed and the pellet washed with 200 µl of cold ethanol (70%) and centrifuged for 5 minutes at 14.000rpm and 4°C. The supernatant is then removed and the pellet air-dried for approximately 20 min (not completely dried). The pellet is dissolved in 100 or 50 µl of ultrapure water. DNA quality and quantity can be measured by gel electrophoresis and absorbance (Nanodrop™). DNA samples are stored at -20°C upon use.

Barcodes and PCR

The most common barcodes for medicinal plants include the following: rbcL, matK, ITS1, and ITS2 (Figure 2.1). However, other barcodes could be also used, as tested by CBOL Plant Working Group (2009, Table 2.1). According to the primers and barcode

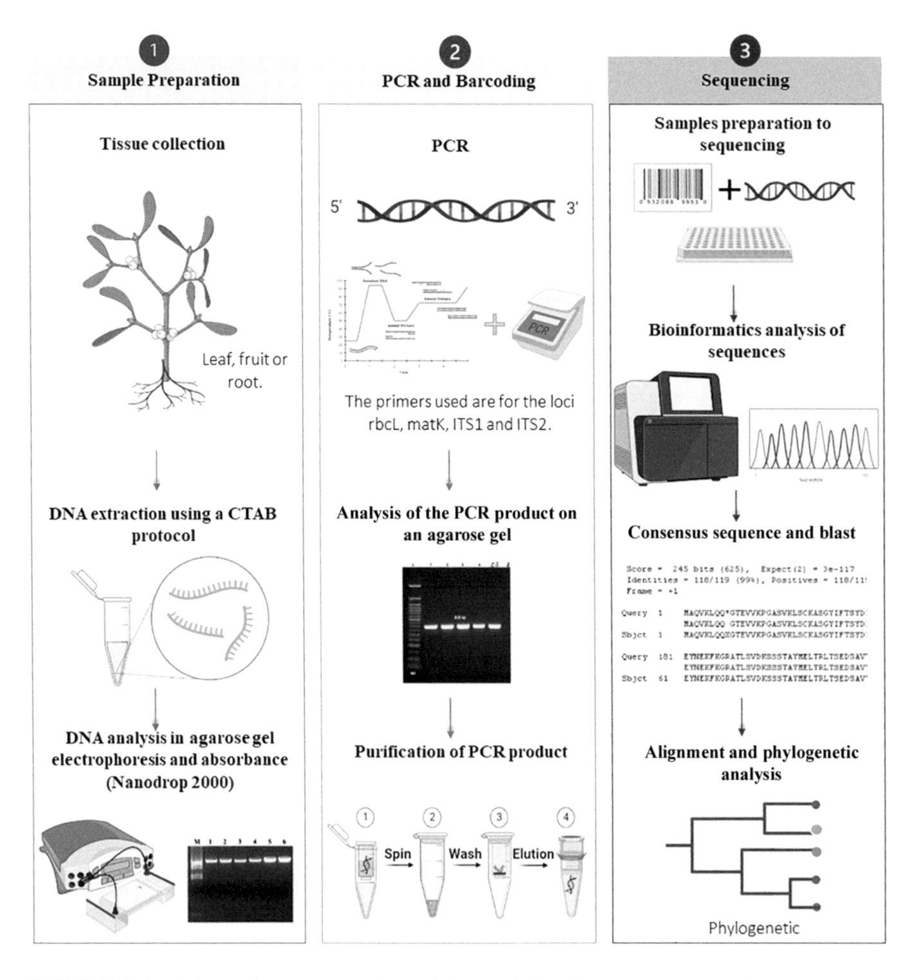

FIGURE 2.1 Schematic representation of the workflow from DNA extraction to phylogenetic analysis using DNA barcodes in plants.

used, different PCR conditions should be applied, especially for the annealing temperature. A master mix could be performed by using the GoTaq® 2X (Promega) according to the manual instructions and 0.5 μM for each primer in a final volume that could range from 10–50 μL. To check for PCR amplifications, a 10 μL PCR volume could be performed. Once amplicons are detected, PCR could be repeated with 30–50 μL volume for sequencing.

PCR condition include a first denaturation step at 95°C for 3 min, followed by 35 cycles of 95°C for 30 sec, annealing temperature (see Table 2.1 for specific barcode primers) for 60 sec, and 72°C for 60 sec; with a final extension step of 72°C for 10 min.

Before sequencing, amplicons are detected by gel electrophoresis (1.5% agarose) by sampling 5 μL. If specific amplicons are detected (clear band, absence of unspecific products or absence of multiple bands), PCR products could be sent to a commercial sequencing facility for PCR purification and Sanger sequencing. At least 2

TABLE 2.1
Primers Used for Amplification of DNA Barcodes for the 7 *Loci* Tested by CBOL Plant Working Group (2009), and ITS1 and ITS2, Which is Used in Several Studies in Medicinal Plants

Locus	Primer pairs	Sequence	Annealing temperature of primers in PCR	Reference
psbA-trnH	trnHf_05	CGCGCATGGTGGATTCACAATCC	60°C	Basak et al. (2019)
	psbA3_f	GTTATGCATGAACGTAATGCTC		
psbK-psbI	psbK_F	TTAGCCTTTGTTTGGCAAG	56°C	Basak et al. (2019)
	psbI_R	AGAGTTTGAGAGTAAGCAT		
rpoB	rpoB_2F	ATGCAACGTCAAGCAGTTCC	50°C	Basak et al. (2019)
	rpoB_3R	CCGTATGTGAAAAGAAGTATA		
rpoC1	rpoC1_2F	GGCAAAGAGGGAAGATTTCG	60°C	Basak et al. (2019)
	rpoC1_4R	CCATAAGCATATCTTGAGTTGG		
atpF-atpH	atpF_F	ACTCGCACACACTCCCTTTCC	56°C	Basak et al. (2019)
	atpH_R	GCTTTTATGGAAGCTTTAACAAT		
*rbc*L	rbcLA_F	ATGTCACCACAAACAGAGACTA AAGC	60°C	Costion et al. (2011)
	rbcLA_R	GTAAAATCAAGTCCACCRCG		
*mat*K	matK_3F_KIMF	CGTACAGTACTTTTGTGTTTA CGAG	56°C	Costion et al. (2011)
	matK_1R_KIMR	ACCCAGTCCATCTGGAAATCTTG GTTC		
ITS1	5a F	CCTTATCATTTAGAGGAAGGAG	56°C	Chen et al. (2010)
	4R	TCCTCCGCTTATTGATATGC		
ITS2	S2F	ATGCGATACTTGGTGTGAAT	56°C	Chen et al. (2010)
	S3R	GACGCTTCTCCAGACTACAAT		

technical replicates (two PCR from same DNA) and at least 3 biological replicates (3 specimens from the same species) should be analyzed.

Bioinformatic Analysis

Sequences are processed using bioinformatics tools, for instance MEGAX (Stecher et al. 2020). Technical replicates should be aligned with MUSCLE and a consensus sequence is generated for each biological replicate using the technical replicates. Consensus sequences are analyzed by BLAST in the GenBank non-redundant nucleotide database (nr). The nr database could contain accessions with the complete plastid genomes or partial sequences corresponding to the barcodes. Therefore, results depend on the sequence availability in the database at the time of the analysis. Accessions are selected for phylogenetic analysis based on BLAST analysis. For phylogenetic analysis for each barcode, the accessions and samples sequences are aligned using MUSCLE and the recommended model from MEGAX is used. The aligned sequences are trimmed at the ends to allow for all sequences to maintain the same range. The maximum likelihood method is performed according to the best model found by MEGAX using a bootstrap test of 1,000 replicates. A scheme with the whole processes from DNA extraction to phylogenetic analysis is indicated (Figure 2.1). The scheme was performed on a medicinal plant collected in Ecuador and identified as *Smilax purhampuy* Ruiz. The phylogenetic analysis of the medicinal plant *Smilax purhampuy* Ruiz collected in Ecuador (Soledispa et al. 2021) is indicated for the ITS2 sequence (Figure 2.2). The *S. purhampuy* Ruiz from Ecuador are

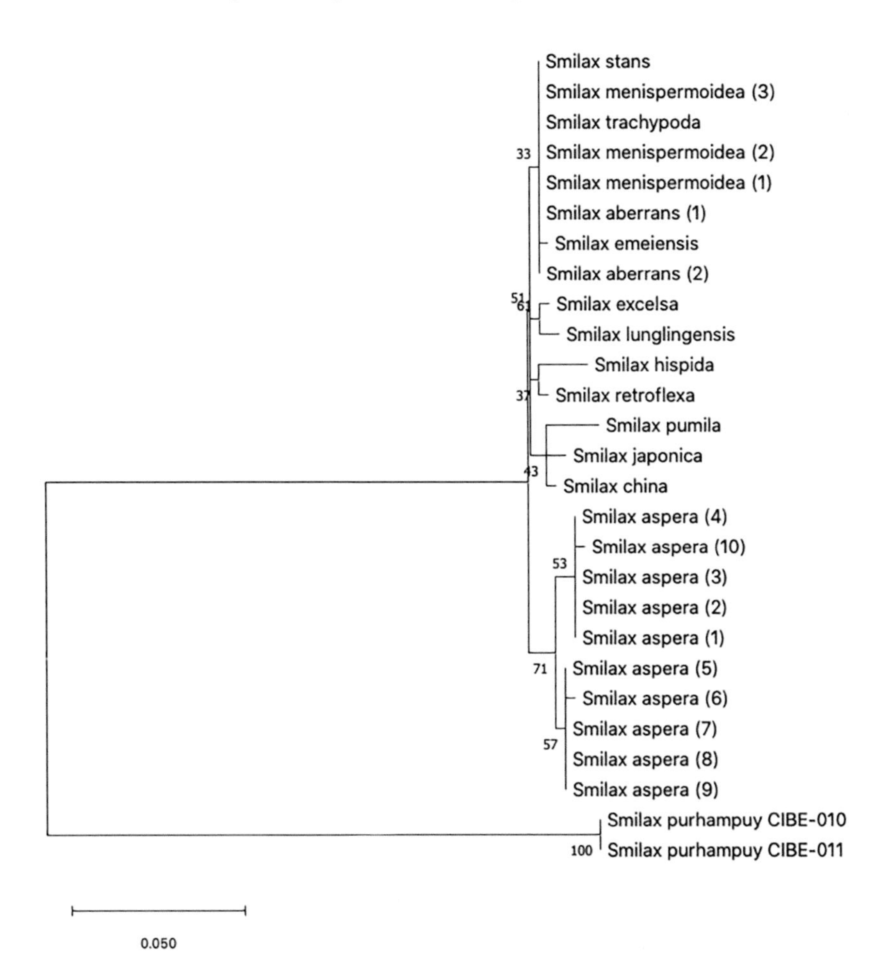

FIGURE 2.2 Phylogenetic analysis of the ITS2 of *Smilax purhampuy* Ruiz from Ecuador with accessions of *Smilax* spp. queried in the GenBank. PCR was performed according to the conditions indicated above and the primers for the amplification of the ITS2 regions is indicated in Table 2.1. Maximum likelihood method based on Tamura 3-parameter model and Gamma distribution. The bootstrap test was performed with 1,000 replicates. The tree is drawn to scale, with branch lengths measured in the number of substitutions per site. Evolutionary analyses were conducted in Mega X (Stecher et al. 2020). *Smilax purhampuy* Ruiz from Ecuador are indicated as CIBE-010 and CIBE-011. An extended analysis of barcode sequences of *S. purhampuy* Ruiz CIBE-010 and CIBE-011 is reported by Soledispa et al. (2021).

grouped in an independent clade when compare from other Smilax species queried in the GenBank. Another specific clade is also indicated for *S. aspera.*

CONCLUSION

DNA barcodes are an important tool as a complement analysis in taxonomic identification of plant species to be used in the medicinal and herbal industry. A combination of different barcodes should be applied to confirm species, which could be dependent on the sequences available in the Genbank. Therefore, it is advisable to apply the analysis to representative species from herbaria or museums, if taxonomic analysis can not be performed. Furthermore, application of DNA barcodes to avoid mislabeling or adulteration in the medicinal and herbal industry is relevant and should be established as a control for consumers and governmental agencies.

REFERENCES

Ajmal Ali, M., Gábor Gyulai, Norbert Hidvégi, Balázs Kerti, Fahad M. A. Al Hemaid, Arun K. Pandey, Joongku Lee. 2014. "The changing epitome of species identification—DNA barcoding". *Saudi J. Biological Sciences* 21, no.3: 204–231.

Armstrong, Kate E., Graham N. Stone, James A. Nicholls, Eugenio Valderrama, Arne A. Anderberg, Jenny Smedmark, Laurent Gautier, Yamama Naciri, Richard Milne, and James E. Richardson. 2014. "Patterns of diversification amongst tropical regions compared: a case study in Sapotaceae." *Frontiers in Genetics* 5: 362.

Basak, Supriyo, Ramesh Aadi Moolam, Ajay Parida, Sudip Mitra, and Latha Rangan. 2019. "Evaluation of rapid molecular diagnostics for differentiating medicinal kaempferia species from its adulterants." *Plant Diversity* 41, no. 3(April): 206–211.

Bustamante, Katherine, Efrén Santos-Ordóñez, Migdalia Miranda, Ricardo Pacheco, Yamilet Gutiérrez, Ramón Scull. 2019. "Morphological and molecular barcode analysis of the medicinal tree Mimusops coriacea (A.DC.) Miq. collected in Ecuador." *PeerJ 7:* e7789.

CBOL Plant Working Group. 2009. "A DNA barcode for land plants." *Proceedings of the National* Academy of Sciences of the United States of America 106, no. 31: 12794–12797.

Chase M.W., et al. 2007. "A proposal for a standardised protocol to barcode all landplants." *Taxon* 56: 295–299.

Chen Shilin, Hui Yao, Jianping Han, Chang Liu, Jingyuan Song, Linchun Shi, Yingjie Zhu, Xinye Ma, Ting Gao, Xiaohui Pang, Kun Luo, Ying Li, Xiwen Li, Xiaocheng Jia, Yulin Lin, Christine Leon. 2010. Validation of the ITS2 region as a novel DNA barcode for identifying medicinal plant species. *PLOS ONE* 5(1): e8613

Costion, Craig, Andrew Ford, Hugh Cross, Darren Crayn, Mark Harrington, and Andrew Lowe. 2011. "Plant DNA Barcodes Can Accurately Estimate Species Richness in Poorly Known Floras." *PLOS ONE* 6, no. 11(November): e26841

Dasmahapatra, K. K., and J. Mallet. 2006. "Taxonomy: DNA barcodes: Recent successes and future prospects." *Heredity* 97, no. 4: 254–255. https://doi.org/10.1038/sj.hdy.6800858

Ford C.S., et al. 2009. "Selection of candidate DNA barcoding regions for use on landplants." *Bot J Linn Soc* 159: 1–11.

Ganie, Showkat Hussain, Priti Upadhyay, Sandip Das, and Maheshwer Prasad Sharma. 2015. "Authentication of medicinal plants by DNA markers." *Plant Gene* 4(December): 83–99.

Ghorbani, Abdolbaset, Barbara Gravendeel, Sugirthini Selliah, Shahin Zarré, and Hugo de Boer. 2017. "DNA barcoding of tuberous orchidoideae: A resource for identification of orchids used in salep." *Molecular Ecology Resources* 17, no. 2: 342–352.

Gonzalez, Mailyn Adriana, Christopher Baraloto, Julien Engel, Scott A. Mori, Pascal Pétronelli, Bernard Riéra, Aurélien Roger, Christophe Thébaud, Jérôme Chave. 2009. "Identification of Amazonian Trees with DNA Barcodes." *PLoS ONE* 4, no. 10(October): e7483.

Hebert, Paul D. N., Alina Cywinska, Shelley L. Ball and Jeremy R. deWaard. 2003. "Biological identifications through DNA barcodes." *Proceedings. Biological sciences* 270, no. 1512: 313–321.

Hollingsworth, Peter M. 2011. "Refining the DNA barcode for land plants." *Proceedings of the National Academy of Sciences* 108, no. 49(December): 19451–19452.

Intharuksa, Aekkhaluck, Yohei Sasaki, Hirokazu Ando, Wannaree Charoensup, Ratchuporn Suksathan, Kittipong Kertsawang, Panee Sirisa-ard, and Masayuki Mikage. 2020. "The combination of ITS2 and PsbA-TrnH Region is powerful DNA barcode markers for authentication of medicinal terminalia plants from Thailand." *Journal of Natural Medicines* 74, no. 1: 282–293.

Kim, W.J., Y. Ji, G. Choi, Y.M. Kang, S. Yang, and B.C. Moon. 2016. "Molecular identification and phylogenetic analysis of important medicinal plant species in genus Paeonia based on rDNA-ITS, matK, and rbcL DNA barcode sequences." *Genetics and Molecular Research* 15, no. 3(August): gmr.15038472.

Kool, Anneleen, Hugo J. de Boer, Åsa Krüger, Anders Rydberg, Abdelaziz Abbad, Lars Björk, and Gary Martin. 2012. "Molecular identification of commercialized medicinal plants in southern Morocco." *PLOS ONE* 7, no. 6(June): e39459.

Kress, J.W., K.J. Wurdack, E.A. Zimmer, L.A. Weigt, and D.H. Janzen. 2005. "Use of DNA barcodesto identify flowering plants." *Proc Natl Acad Sci USA* 102: 8369–8374.3.

Kress, W. John, and David L. Erickson. 2007. "A two-locus global DNA barcode for land plants: the coding rbcL gene complements the non-coding trnH-psbA spacer region." *PLoS One* 2: e508.

Kumari, Rajesh, and Mita Kotecha. 2016. "A review on the standardization of herbal medicines." *International Journal of Pharma Sciences and Research* 7, no. 2 (February): 97–106.

Li, De-Zhu, Lian-Ming Gao, Hong-Tao Li, Hong Wang, Xue-Jun Ge, Jian-Quan Liu, Zhi-Duan Chen, Shi-Liang Zhou, Shi-Lin Chen, Jun-Bo Yang, Cheng-Xin Fu, Chun-Xia Zeng, Hai-Fei Yan, Ying-Jie Zhu, Yong-Shuai Sun, Si-Yun Chen, Lei Zhao, Kun Wang, Tuo Yang, and Guang-Wen Duan. 2011. "Comparative analysis of a large dataset indicates that internal transcribed spacer (ITS) should be incorporated into the core barcode for seed plants." *Proceedings of the National Academy of Sciences* 108, no. 49(December): 19641–19646.

Li, Xian-kuan, Bing Wang, Rong-chun Han, Yan-chao Zheng, Hai-bo Yin Yin, Liang Xu, Jian-kui Zhang, Bao-li Xu. 2013. "Identification of medicinal plant Schisandra chinensis using a potential DNA barcode ITS2." *Acta Societatis Botanicorum Poloniae* 82, no. 4: 283–288

Liu, Juan, Hai-Fei Yan, Steven G Newmaster, Nancai Pei, Subramanyam Ragupathy, and Xue-Jun Ge. 2015. "The use of DNA barcoding as a tool for the conservation biogeography of subtropical forests in China." *Diversity and Distributions* 21, no. 2(February): 188–199.

Mishra, Priyanka, Amit Kumar, Akshitha Nagireddy, Daya N Mani, Ashutosh K Shukla, Rakesh Tiwari, and Velusamy Sundaresan. 2016. "DNA Barcoding: An Efficient Tool to Overcome Authentication Challenges in the Herbal Market." *Plant Biotechnology Journal* 14, no. 1(January): 8–21.

Purushothaman, Natarajan, Steven G. Newmaster, Subramanyam, Ragupathy, Nithaniyal, Stalin, Damodaran, Suresh, Duraipandian R. Arunraj, Gunadayalan, Gnanasekaran, Sophie L. Vassou, Dindigal, Narasimhan, and Madasamy Parani. 2014. "A tiered barcode authentication tool to differentiate medicinal Cassia species in India." *Genetics and Molecular Research* 13, no. 2(April): 2959–2968.

Richardson, James E., Azrul M. Bakar, James Tosh, Kate Armstrong, Jenny Smedmark, Arne A. Anderberg, Ferry Slik, and Peter Wilkie. 2013. "The influence of tectonics, sea-level changes and dispersal on migration and diversification of Isonandreae (Sapotaceae)." *Botanical Journal of the Linnean Society* 174, no. 1: 130–140.

Sarmiento-Tomalá, Glenda, Efrén Santos-Ordóñez, Migdalia Miranda-Martínez, Ricardo Pacheco-Coello, Ramón Scull- Lizama, Yamilet Gutiérrez-Gaitén, René Delgado-Hernández. 2020. "Short communication: molecular barcode and morphology analysis of Malva pseudolavatera Webb & Berthel and Malva sylvestris L from Ecuador." *Biodiversitas* 21, no. 8(August): 3554–3561.

Sarwat, Maryam, and Manu Mayank Yamdagni. 2016. "DNA barcoding, microarrays and next generation sequencing: recent tools for genetic diversity estimation and authentication of medicinal plants." *Critical Reviews in Biotechnology* 36, no. 2: 191–203.

Sass, Chodon, Damon P. Little, Dennis Wm Stevenson, and Chelsea D Specht. 2007. "DNA barcoding in the cycadales: testing the potential of proposed barcoding markers for species identification of cycads." *PLOS ONE* 2(November): e1154.

Soledispa, Pilar, Efrén Santos-Ordóñez, Migdalia Miranda, Ricardo Pacheco, Yamilet Irene Gutiérrez Gaiten, Ramón Scull. 2021. "Molecular barcode and morphological analysis of *Smilax purhampuy* Ruiz, Ecuador." *PeerJ* 9(March): e11028.

Srivastava, Swati, Sanchita Mili Bhargava, and Ashok Sharma. 2016. "Barcoding of medicinal plants." In: Tsay HS., Shyur LF., Agrawal D., Wu YC., Wang SY. (eds) *Medicinal Plants—Recent Advances in Research and Development*. Springer, Singapore.

Stecher, Glen, Koichiro Tamura, and Sudhir Kumar. 2020. "Molecular evolutionary genetics analysis (MEGA) for MacOS." *Molecular Biology and Evolution* 37, no. 4(April): 1237–1239.

Techen, Natascha, Parveen Iffat, Pan Zhiqiang, and Ikhlas AKhan. 2014. "DNA barcoding of medicinal plant material for identification." *Current Opinion in Biotechnology* 25(February): 103–110.

Tripathi, Abhinandan Mani, Antariksh Tyagi, Anoop Kumar, Akanksha Singh, Shivani Singh, Lal Babu Chaudhary, and Sribash Roy. 2013. The internal transcribed spacer (ITS) region and trnhH-psbA are suitable candidate loci for DNA barcoding of tropical tree species of India. *PLOS ONE* 8(February): e57934.

Urumarudappa, Santhosh Kumar J, Chayapol Tungphatthong, and Suchada Sukrong. 2019. "Mitigating the impact of admixtures in Thai herbal products." *Frontiers in Pharmacology* 10(October): 1205.

Urumarudappa, Santhosh Kumar Jayanthinagar, Navdeep Gogna, Steven G Newmaster, Krishna Venkatarangaiah, Ragupathy Subramanyam, Seethapathy Gopalakrishnan Saroja, Ravikanth Gudasalamani, Kavita Dorai, and Uma Shaanker Ramanan. 2016. "DNA barcoding and NMR spectroscopy-based assessment of species adulteration in the raw herbal trade of saraca asoca (Roxb.) Willd, an important medicinal plant." *International Journal of Legal Medicine* 130, no. 6(September): 1457–1470.

Vivas, Caio Vinicius, Ramiris César Souza Moraes, Anderson Alves-Araújo, Marccus Alves, Eduardo Mariano-Neto, Cássio van den Berg, and Fernanda Amato Gaiotto. 2014. "DNA barcoding in atlantic forest plants: What is the best marker for sapotaceae species identification?" *Genetics and Molecular Biology* 37, no. 4(October): 662–670.

Xiang, Xiao-Guo, Jing-Bo Zhang, An-Ming LU, and Rui-Qi LI. 2011. "Molecular identification of species in juglandaceae: A tiered method." *Journal of Systematics and Evolution* 49, no. 3(May): 252–260.

Yu, Jie, Xi Wu, Chang Liu, Steve Newmaster, Subramanyam Ragupathy, and W John Kress. 2021. "Progress in the use of DNA barcodes in the identification and classification of medicinal plants." *Ecotoxicology and Environmental Safety* 208: 111691.

Zhang, Dequan, Bei Jiang, Lizhen Duan, Nong Zhou. 2016. "Internal transcribed spacer (ITS), an ideal DNA barcode for species discrimination in Crawfurdia Wall. (Gentianaceae)." *African Journal of Traditional, Complementary and Alternative Medicines* 13, no. 6: 101–106.

Zhou, Jing, Wencai Wang, Mengqi Liu, and Zhenwen Liu. 2014. "Molecular authentication of the traditional medicinal plant peucedanum praeruptorum and its substitutes and adulterants by DNA—barcoding technique." *Pharmacognosy Magazine* 10, no. 40(October): 385–390.

3 Pharmacological Studies as Essential Tools for the Scientific Validation of the Traditional Use of Herbal Drugs

Some Considerations

René Delgado Hernández, Alejandro Felipe González, Bárbara Beatriz Garrido Suarez and Idania Rodeiro Guerra

CONTENTS

INTRODUCTION

Over time natural products have been used as traditional medicines, remedies, potions and oils without any knowledge about the bioactive compounds but relying on results of hundreds of centuries of empirical knowledge (Bernardini et al., 2018). It is known that around 20–30% of the medicines available on the market are derived

DOI: 10.1201/9781003173991-4

from natural products (Majouli et al., 2017). Some natural product-derived drugs that are a hallmark of modern pharmaceutical care include quinine, theophylline, penicillin G, morphine, paclitaxel, digoxin, vincristine, doxorubicin, cyclosporine and vitamin A, among many other examples (Mubanga et al., 2017). Therefore, traditional knowledge about medicinal plants is the first clinical evidence on efficacy of herbal medicine; however, scientific studies are necessary to corroborate ethnobotanical information (Calixto, 2000).

Current drug discovery strategies and modern medicine discard the use of whole plant extracts and are driven by single compound-based medicine. Using the whole plant with no isolation of components, as practiced in traditional medicine, produces a better therapeutic effect than individual compounds. This is important, as most of the plant metabolites likely work in a synergistic fashion or concurrently to give the plant its therapeutic effect. However, searching for new drug candidates from natural products is often made difficult by the complexity of molecular mixtures (Thomford et al., 2018).

There is a close relationship between the chemical composition and the biological activity of natural products where the pharmacological efficacy of herbal medicines will depend on their chemical composition (Ernst, 2005). For this reason, it is necessary to identify chemical markers in order to guarantee their quality, efficacy and safety (Butt et al., 2018). According to the European Medicines Agency (EMA), chemical markers are "chemically defined constituents or groups of constituents of an herbal substance, an herbal preparation or an herbal medicinal product, which serve for quality control purposes, independent of whether they have any therapeutic activity". EMA describes two different categories of markers: the constituents of an herbal medicine responsible for its therapeutic activity or active markers; and the constituents that are characteristic of its taxon or analytical markers (Rivera et al., 2017). Recent trends led to identify active markers for the quality control of natural products in order to guarantee better efficacy of the final product, however, for most natural products the therapeutic components have not been fully elucidated or sufficiently monitored (Li et al., 2008).

In this context, the pharmacological studies play an essential role for the characterization of natural products in order to demonstrate their efficacy and safety. In fact, the introduction of *in vitro* pharmacological analysis in the parameters of the quality control of natural products can be an interesting point of view to consider. The present review discusses the importance of pharmacology and toxicology analysis in natural products, and the new trends in the discovery of therapeutic drugs derived from natural products.

FROM ETHNOMEDICINE TO PHARMACOLOGICAL STUDIES

It is known that biological activities of plants depend on their chemical composition (Ernst, 2005). On the other hand, the main problems associated with herbal medicine research are the complexity chemical composition as well as the variability of factors such as climatic conditions, soil characteristics, age of the plant, phenological stages and genetic variability (Li et al., 2020). Therefore, searching for new drugs requires advanced methods and a multidisciplinary team of researchers, including botanists, pharmacognosists and pharmacologists.

Pharmacognosy is the science that studies the physical, chemical, biochemical and biological properties of drugs, drug substances, or potential drugs of natural origin, and it also includes the search for new drugs from natural sources (Biswas, 2015). This definition includes ethnomedical, phytochemical and pharmacological studies, as well as quality control analysis of natural products (Cahlíková et al., 2020).

Ethnomedicine is the study of traditional systems of medicine practiced by various ethnic groups or communities around the world to cure a particular disease, especially by indigenous peoples of a particular locality. It comprises studies of traditional medicine and its methods as well as anthropological aspects (Disco et al., 2020).

Traditional knowledge about medicinal plants is the first clinical evidence on the efficacy of herbal medicines (Calixto, 2000). This knowledge was empirically acquired and passed down through legends, pictographs and various monographs to the present day (Petrovska, 2012). Ethnomedical studies have led to information about the traditional use of natural products, which is used as primary evidence for the development of new natural drugs.

There is a strength in relationship between traditional and modern medicine in the field of natural products. Traditional medicine has been practiced for many years, but it is based on empirical experiences acquired by observational methods. Modern medicine explains the traditional knowledge of herbal medicine based on scientific information, mainly pharmacognostic and pharmacological studies (Petrovska, 2012). The integration of traditional and modern medicine may allow obtaining new drugs for the treatment of several diseases and to validate ethnomedical information, all to improve public health and to facilitate access to drugs at lower costs (Figure 3.1).

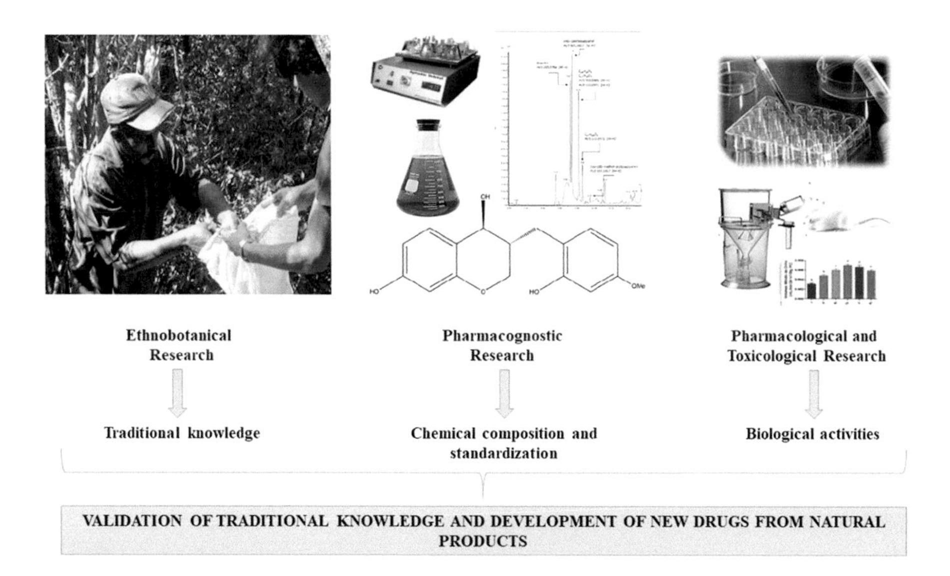

FIGURE 3.1 Integration of traditional and modern medicine to obtain new drugs.

Phytochemistry is defined as the chemical profile of medicinal herbs and an understanding of analytical tests for identification of the herbs and for a quantitative assessment of any known active ingredients. In this way, phytochemical studies have two main objectives: to isolate new bioactive compounds for obtaining synthetic or semi-synthetic drugs and to identify chemical markers for the quality control of herbal products (Jin et al., 2010; Phillipson, 2007).

Some of the most used drugs in current therapies have their origins in natural products. For example: aspirin (*Salix alba*), morphine (*Papaver somniferum*), digoxin (*Digitalis purpurea*), reserpine (*Rauwolfia serpentine*), vincristine and taxol (*Catharanthus rosea*). Most of these plant-derived drugs were originally discovered through the study of traditional uses and folk information of indigenous people, and some of these could not be substituted despite the enormous recent advancements of synthetic chemistry (Aftab, 2018). Other research has focused on the identification of bioactive markers for the quality control of herbal medicine (Table 3.1). Both trends, new bioactive compounds for obtaining drugs and bioactive markers for the quality control of herbal medicine, require pharmacological studies to evaluate their biological properties.

Pharmacology is the study of the mechanisms and chemical effects of compounds on living organisms (pharmacodynamic), and the effects of the living organisms on the chemicals, including absorption, distribution, metabolism and excretion (pharmacokinetic). Also, pharmacology includes the effect of the drugs at high dosages (toxicology) (Hacker, 2009). Therefore, pharmacological studies evaluate and explain

TABLE 3.1
Some Chemical Markers Identified in Medicinal Plants

Specie	Chemical marker	Reference
Caesalpinia spp.	Homoisoflavonoids	Baldim et al. (2017)
Cecropia spp.	chlorogenic acid, flavonoids and triterpenoids	Rivera et al. (2017)
Corydalis mucronífera	sanguinarine	Liu et al. (2017)
Evodiae rutecarpa	evodiamine and rutaecarpine	Kim et al. (2010)
Jatrorrhiza palmate	tetrahydrocolumbamine	Liu et al. (2017)
Lantana spp.	essential oils (β-caryophyllene, β-phellandrene, elixene)	Sena et al. (2010)
Lycoris radiata	galanthamine and ungerimine	Shi et al. (2014)
Murraya spp.	polymethoxylated flavonoids	Liang et al. (2020)
Ouratea spp.	biflavonoids (amentoflavone, agathisflavone)	Fidelis et al. (2014)
Salvia miltiorrhiza	cryptotanshione, trijuganone B, 15,16-dihydrotanshinone I)	Liang et al. (2017)
Schisandrae chinensis	gomisin A, gomisin N and schisandrin	Kim et al. (2010)
Waltheria spp.	benzamide derivates (waltherine-A, methyltembamide)	Caridade et al. (2018)
Xiphidium spp.	phenylphenalenone and related compounds	Optiz et al. (2002)
Mangifera indica	mangiferin	Núñez et al. (2002)
Connarus venezuelanus	rapanone	de la Vega et al. (2017)

the biological effects of herbal medicine and natural products through pre-clinical and clinical studies.

Herbal medicine involves multiple pathways to achieve pharmacological effects due to the complex chemical composition of plants. This is the first problem in pharmacological studies because molecular targets for herbal medicine are unknown. For this reason, the pharmacological studies in natural products require new methods and technologies for the elucidation of molecular mechanisms (Atanasov et al., 2021). In this context, the study of the mechanisms of the bioactive markers by means of traditional assay-based methods is an acceptable solution for this problem; but it is a costly and time-consuming process. Therefore, the application of *in silico* methods would be an alternative method to search drug-targets for medicinal plant ingredients, however, *in vivo* studies are necessary as final confirmation (Chen et al., 2003).

Summarizing, the combination of ethnomedical, phytochemical, toxicological and pharmacological studies are a key step to validate traditional knowledge on medicinal plants. Ethnomedical studies provide information about traditional use of medicinal plants, while phytochemical and pharmacological studies establish the scientific basis through the relationship between chemical composition and biological properties (Xie et al., 2018). Finally, the standardization of herbal medicine will guarantee the reproducibility of the results, taking into consideration chemical variability to internal and external factors. Thus, it will be possible to obtain a final product with the security and efficacy required for the treatment of several diseases (Figure 3.1).

PHARMACOLOGICAL MODELS USED IN THE ANALYSIS OF HERBAL MEDICINE. BIOETHICAL CONSIDERATIONS

As mentioned above, pharmacological studies are a necessary step for the validation of traditional knowledge in medicinal plants and to evaluate the safety and efficacy of new drugs based on natural products (Atanasov et al., 2021).

Traditionally, *in vivo, in situ* and *in vitro* pharmacological models have been used for the analysis of extracts, fractions and biologically active compounds. These models require the use of animal labs or isolated cells, tissues or organs of animal or humans (Lahlou, 2007). However, new trends are reducing the use of animal labs in pharmacological research because many people agree that it is a cruel and inhumane practice (Prasad, 2016). For this reason, many important agencies have regulated the use of animal labs and have proposed the replacement, refinement and reduction of animal experiments. Therefore, alternative methods are required for the research of new molecules and extracts in pharmacological studies (Badyal and Desai, 2014).

Replacement is divided into absolute and relative, independent of the total or partial substitution of animal labs in the experiments. Absolute replacement implies the non-use of laboratory animals in the experiments. For example, *in silico* computer modeling and *in vitro* methodologies are based on tissue engineering and chemical analysis. Another possibility is relative replacement, which avoids or replaces the use of protected animals. These methods include established animal cell lines, animal cells, tissues and organs collected from animals sacrificed by a human technique, abattoir material and use of invertebrates, larval forms of amphibians, fish, bacterial, fungi and others (Badyal and Desai, 2014). In this way, several experiments have been reported in scientific literature (Table 3.2). Despite that, there are not good alternatives to animal testing because alternatives do not offer sufficient pharmacological information about safety and efficacy of the products (Prasad, 2016).

Reduction and refinement take into account the use of animal labs; they minimize animal use and enable researchers to obtain comparable levels of information from fewer animals, or allow them to obtain more information from the same number of animals and to decrease invasive methods during the experiments in animal labs (Badyal and Desai., 2014). Table 3.3 lists some biological experiments in animal models applied to different pharmacological activities.

TABLE 3.2
Examples of Alternative Methods Applied to Pharmacological Research

Activity	Pharmacological Model	Reference
***1. In silico* analysis based in computational software (Total replacement)**		
Antiviral (Covid-19)	CODESSA software (QSAR)	Si et al. (2021)
Antibacterial (*E. coli*)	BMDP software (QSAR)	Suay et al. (2020)
***2. In vitro* analysis based in chemical solutions (Total replacement)**		
Antioxidant	DPPH, FRAP, ABTS	Yashin et al. (2017)
Antilithiatic	Inhibition of CaOx crystallization	Bensatal et al. (2020)
***3. In vitro* models in cells and other organisms and microorganisms (Relative replacement)**		
Analysis of Cell Viability	MTT viability Assay	Kumar et al. (2018)
Antiproliferative Activity	Human tumor cells line	Qiong et al. (2020)
Antimicrobial	Agar well diffusion in plates	Manandhar et al. (2019)

QSAR: Quantitative Structure Activity Relationships; DPPH: 1,1-diphenyl-2-picrylhydrazyl; FRAP: Ferric Reducing Activity Power; ABTS: 2,2'-azino-bis (3-ethylbezothiazoline-6-sulfonic acid) CaOx: calcium oxalate; MTT: 3-(4,5-dimethylthiozol-2-yl-2,5-diphenyltetrazolium bromide

TABLE 3.3
Example of Some Pharmacological Activities Using Different Methods of Evaluation in Animal Models

Pharmacological activity	Animal model	Reference	Observations
Analgesic activity	Formalin test	Dubuisson and Dennis (1977) Okuda et al. (2001)	Formalin-induced biphasic nocisensitive behavioral response: phase I (early, nociceptive) and phase II (late, tonic). A persistent pain models.
	Chronic constriction injury model (CCI)	Bennett and Xie. (1988)	A model of peripheral neuropathic pain. CCI is a compressive, ischemic, inflammatory neuropathy of the sciatic nerve.

Pharmacological activity	Animal model	Reference	Observations
	Chronic post ischemic pain (CPIP)	Coderre et al. (2004) Xanthos et al. (2008) Coderre and Bennett (2010) Millecamps et al., (2010)	A model of complex regional pain syndrome type 1 (CRPS-I) induced by ischemia and reperfusion injury of the left hind paw.
	Behavioral test of mechano-allodynia (to measure abnormalities of stimulus-evoked pain in chronic pain models)	Chaplan et al. (1994)	Quantitative assessment of tactile allodynia in the rat hind paw by measuring the 50% withdrawal response to stimulation with von Frey filaments.
Motor coordination Nervous system evaluation	Rota-rod test	Rosland et al. (1990)	To evaluate possible nonspecific effects on peripheral and nervous system coordination such as muscle relaxation or sedation.
Angiogenesis (Neovascularization)	Chicken egg chorioallantoic membrane (CAM) angiogenesis assay	Attoub et al. (2013)	Use briefly and fertilized eggs.
Tumor angiogenesis	Neovascularization in melanoma syngeneic model of tumor	Yeh et al. (2001)	Neovascularization and metastasis induction by B16F10 melanoma cell line.
Inflammation	Neutrophil migration to peritoneal cavity	Dal Secco et al. (2006)	Inflammatory stimuli by i.p. injection of carrageenan in naive mice.
	Paw edema	Morris Ch J. (2003)	Inflammatory stimuli carrageenan.
	Vascular permeability assay	Silva et al. (2013)	Inflammatory stimuli carrageenan and Evans blue were used for measuring vascular extravasation.
	acute lung injury model	Chen et al. (2010)	Lipopolysaccharide-induced acute lung injury model in respiratory medicine.

The emergence of new diseases that require treatment, where medicinal plants have been shown to be effective through traditional knowledge, requires an increase in efforts to find new drugs based on natural products. In this context, modern pharmacology will require combining traditional and modern methodologies in order to optimize pharmacological studies and to obtain all necessary information for development of new herbal medicines. It is important to use all available tools in order to obtain the best results in a minimal time, guaranteeing the adequate use of animal labs.

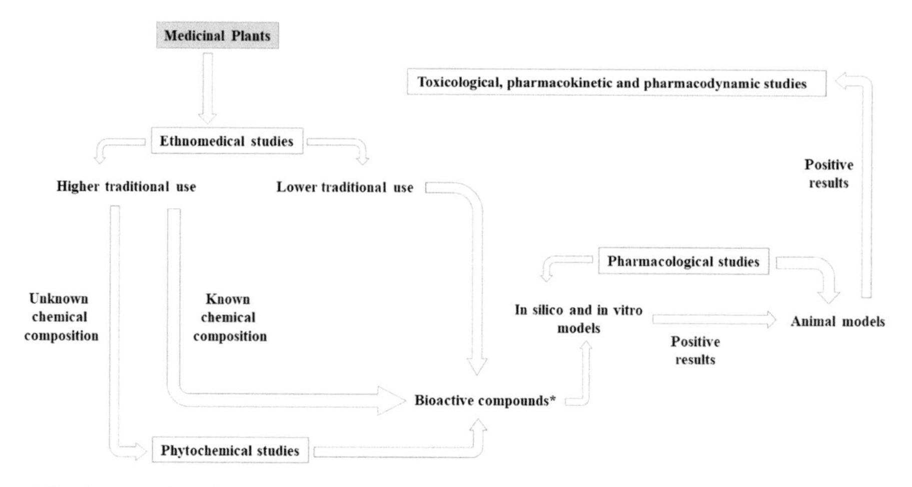

FIGURE 3.2 Methodological proposal of pharmacological research in natural products in keeping with current trends.

According to previous discussion, good methodology for pharmacological screening would include an adequate use of the available information to choose potential drugs with future therapeutic applications and to reduce the use of animal labs in the experiments. In natural products, there are two main sources of information: traditional knowledge and bioactive compounds isolated. Therefore, a good selection of vegetal material will be related to the frequency of traditional use demonstrated through ethnomedical studies, the presence of biologically active compounds identified from natural resources in highest concentrations, or the combination of both. A second step will be to develop *in silico* and *in vitro* models (where possible) in order to choose the most pharmacologically active ingredients (total extracts, fractions, isolated compounds). This step will reduce experiments in animal models, the experimental time and research costs. Finally, the active ingredients would be evaluated in animal and molecular models in order to demonstrate the pharmacological activity, mechanism of action, pharmacokinetic parameters and toxicological information. Figure 3.2 showed a tentative graphical representation of the general proposal methodology for the study of drug candidates from natural sources.

SOME CONSIDERATIONS ABOUT THE NON-CLINICAL PHARMACOLOGICAL STUDIES

During the development of non-clinical pharmacological studies, it is very important to comply strictly with the main international regulations established for this activity; for example, Europe's regulations on animal protection (Directive 86/609) (EMEA, 2006; Louhimies, 2002), Declaration of Helsinki and the Guide for the Care and Use of Laboratory Animals as adopted and promulgated by the US National Institute of

Health (NIH Publication No. 85–23, revised 1996). Regardless of these reference regulations, each country must adopt its own laws and regulations, complying with the main international agreements in this field. Despite that, it is very important that different experimental protocols are approved and reviewed by the appropriate Ethics Committees for Animal Experimentation of each institution.

Each pharmacological research should follow certain percepts: 1) the design of well-structured dose-response studies with the use of adequate controls, reference drugs according to the pharmacological activity studied, among others, and 2) a detailed study of pharmacological efficacy over time, to assess whether the product under study has preventive or therapeutic actions (or both) and its temporary durability. In these studies, it is very appropriate to design pharmacometric investigations to analyze the absorption, distribution, metabolism and excretion processes of the product under study and pharmacodynamic studies to evaluate potential toxicity in vital functions relative to time and intensity of drug exposure (Figure 3.3) (Pugsley et al., 2008).

In this sense, the core studies should be oriented to explore the activity of the product under study in essential pharmacodynamic actions, it being most advisable to develop different investigations in established experimental models to explore possible actions in the cardiovascular system, respiratory system and nervous system; and in a complementary way, develop other studies in experimental models that explore the possible action of the product under study in the gastrointestinal and renal systems.

For the pharmacological characterization of drug candidates, the exploration of the main mechanisms of action that characterize the pharmacological efficacy of the products under study acquires special relevance. In this sense, it is very important to have certified laboratories equipped with the main equipment; as well as with the most modern experimental methodologies and trained scientific staff with the essential skills for this type of activity of research-development of new drugs; all within an adequate quality control system and with the corresponding validation of its experimental processes. For this purpose, the use of cell cultures for *in vitro*

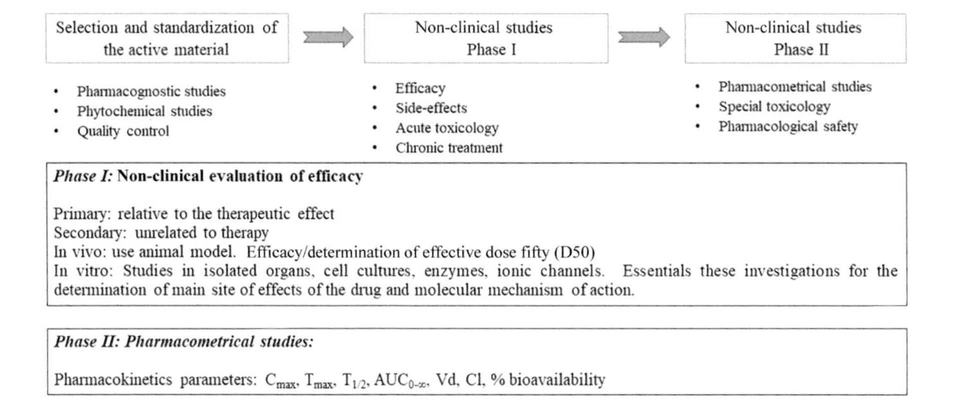

FIGURE 3.3 General design of non-clinical pharmacological studies.

studies is essential, using appropriate and certified cell lines, or primary cultures, or isolated organs, among other methodologies that guarantee excellence and depth in the studies carried out; as well as their corresponding *in vivo* pharmacological models, according to the type of pharmacological activity that is being explored, with the reasoned implementation of rationality in the use of "*in vivo* models", in order to arrive at the possible mechanism of action of the products under study, and finally have true "candidates for new drugs" that may later be suitable, according to the research carried out, for the possible health registration process by the regulatory entities of the new drugs, once they comply with the all phases and stages established for this purpose. (US Food and Drug Administration, 2005).

SOME GENERAL ASPECTS TO CONSIDER FOR *IN VIVO* PHARMACOLOGICAL STUDIES

Experimental animals: The requirement is to select the most appropriate animal species (rodents, rabbits, guinea pigs, dogs, and others) under strict ethical control and complying with the 3R regulations: reduce, refine and replace. In all cases, compliance with international regulations concerns the care and maintenance of the selected species, conditions of the animal house, temperature controls, humidity, light cycles, adequate microbiological control, as well as the training of technicians and professionals involved in animal experimentation. Therefore, the development of qualification courses for technicians and researchers, and the granting of licenses and rigorous control and inspection systems by competent regulatory bodies, are essential to achieve the success of this activity. In this sense, animal experimentation areas must be certified and approved for this activity by the regulatory bodies and agencies established for this purpose in each country; FDA (USA), EMEA (Europe), ANVISA (Brazil), INVIMA (Colombia), CECMED (Cuba), among others.

Product under study: Adequate chemical and analytical characterization is essential. A product's solubility should be known in different solvents that are appropriate for its administration in selected experimental animals. The most appropriate and ideal would be that the products were soluble in water; if this is not fulfilled, it could be explored with some surface-active substances that can help to solubilize the product. It is also possible to use dimethylsulfoxide (DMSO) or solutions with ethanol or other alcohols, as well as some organic solvents; always at adequate concentrations, which are not toxic to the animal and which do not interfere with the pharmacological action that is being explored. In all cases, the vehicle that is used to solubilize the product should be used in the experimental design that is developed, as one more product in the study, and enable an experimental group with, or control with the vehicle, at the same concentration in which it is found in the product under study.

Dose Design: At least 3 doses of the product under study must be used (geometrical/log scale). To establish these doses, toxicological studies must have been previously carried out, to know the safety margin of the product, which will allow to know the therapeutic index of the product. In these studies, the first dose of No Observed Adverse Effect Level (NOAEL) and the effective dose 50 (ED50) therapeutic index (ti) = NOAEL dose/effective dose should be determined. It is important to establish the equivalence factors between the doses used in animals and their extrapolation for use in humans; use with adjustments to the body surface. A general equivalence range has been established. (For more information, observe Table 3.1 in Reference Nair and Jacob, 2016.)

Administration route: Another important aspect is the route of administration. This should be similar to the one later intended to be used during the application of the product to humans. First, the oral route should always be explored (by gavage, POS), with the exception of products that are developed for topical purposes. Other routes to explore are the subcutaneous (sc), intramuscular (im), and intravenous (iv). In non-clinical studies using rodents, the intraperitoneal (ip) route is used regularly.

Establishment of Experimental groups: At least 3 experimental groups should be considered in pharmacological studies: 1) Control groups (vehicle—tap water, saline solution and others); 2) Experimental group (treated with extract/fraction or isolated compound) with at least 3 doses (geometrical/log scale), and 3) Positive control group (standard drug with a known action according to pharmacological activity in study). At the end of the study, it is very important to establish the appropriate dose and time course-response curves (specificity of action).

In general, when starting the study of a product as a candidate for a future drug, it was initially recommended as a strategy for the development of *in vivo* pharmacological studies to first carry out observational studies, based on the Hippocratic test or Irwin Test. The Irwin observational test has been commonly used to assess the effects of a new substance on behavior and essential physiological functions. The results of the Irwin test have been used to determine the potential toxicity and to select the appropriate doses to use the products studied in a specific therapeutic activity. The Irwin test can also be used in a safety approach to detect adverse effects of a new compound on general behavior and to assess its acute neurotoxicity. In particular, the data obtained from the Irwin test can help determine the dose range that will be used in other safety tests. Furthermore, the Irwin test can provide a first and pertinent orientation towards a specific therapeutic indication, a specific mechanism of action or a specific physiological function (Mathiasen and Moser, 2018).

However, at present, non-clinical toxicological studies according to methodologies established at an international level by regulatory entities authorized for this purpose, and the design of appropriate pharmacological studies that are integrated into the so-called "safety pharmacology", as well as the development of *in vitro* and *in vivo* pharmacological studies, aimed at conceptually demonstrating the properties intended to be determined for a given natural product, have played a decisive role, which go beyond the traditional so-called "Irwin tests". Above all, it is about obtaining the necessary non-clinical evidence that demonstrates the medicinal properties of the products used by the so-called "traditional medicine" and in this way achieve that ethnomedicine as a science, accompanied by pharmacognosy, analytical chemistry, toxicology and experimental pharmacology, with the opportune use of computer science and other modern tools; allow to provide the necessary scientificity that should accompany traditional therapeutic methods. In this sense, a general design is carried out for the development of non-clinical pharmacological studies *in vivo*, in particular with herbal extracts or isolated compounds of these; it could follow the following diagrams (Figures 3.2 and 3.3) and provide new phytotherapeutic products, in addition to validating those that are traditionally already employed by the population.

ROLE OF TOXICOLOGICAL STUDIES IN THE EVALUATION OF SAFETY AND EFFICACY OF NATURAL PRODUCTS

Despite the growing global demand of herbal medicine, there are still concerns associated with not only their use, but their safety (WHO, 2019). Herbal medicines are generally referred as safe and are presented as being "natural" and completely "safe" due to long history of use of the natural products. Meanwhile, today the necessity to conduct toxicity studies of herbal products is broadly recognized. Although there are scientific evidences, most of the plants have not undergone exhaustive toxicological tests, such as are required for modern pharmaceutical compounds. Thus, within the requirements for the non-clinical evaluation of drugs of natural origin, the toxicological tests are included (Ifeoma and Oluwakanyinsola, 2013; Aydin et al., 2016).

Similar to synthetic compounds, a battery of toxicity tests must be conducted to provide information from the intrinsic toxicity of the natural product being evaluated up to the identification of possible toxic effects at the cellular, molecular and organism levels. For the evaluation of toxicity potential of a natural entity, all available traditional information should be considered prior to conducting the toxicological studies. The evaluation includes the general toxicology assays (single- and repeat-dose studies) and special toxicity studies such as genotoxicity assays, which permit to assess the potential for induction of genetic mutations or chromosomal damage, as well as reproductive toxicity assays to evaluate effects on development of the zygote, safety pharmacology evaluation to know about potential adverse effects of the product on vital organs (e.g., cardiovascular, central nervous, and respiratory systems) and the conduction of long-term rodent studies, as in the case of carcinogenic studies (OECD, 2008b; FDA, 2010, 2011). In addition, other studies, such as phototoxicity, neurotoxicity, cardiotoxicity and immune toxicity testing, toxicokinetic information and drug interaction studies must be also conducted (ICH, 2000; Ifeoma and Oluwakanyinsola, 2013).

The conditions under which the toxicological studies must be carried out were set according to the criteria of experts from different regulatory agencies. For example, for the single dose studies the determination of the LD_{50} is not necessary to register a new product. Instead, it is recommended to use the Fixed Dose Method, the Acute Toxicity Class Method or the Up and Down Method, assays which the 3Rs principles (reducing the number of animals used in each test, refining existing procedures, reducing stress and their suffering) are considered (OECD, 2002a, 2002b, 2022). In this way, current regulations compel the use of *in vitro* tests as part of the toxicology battery of tests.

In vitro toxicity studies allow the detection of alterations in basic cellular functions due to exposure to a compound (O'Brien and Haskings, 2006). Within *in vitro* tests, cytotoxicity tests stand out. In this case, it is recommended to carry out at least two tests to measure different damage criteria. The tests must be based on the fact that a cytotoxic compound affects at least one or more processes involved in cell proliferation, such as: DNA synthesis, organelle function, cell membrane integrity or cell synthesis proteins.

As regards *in vivo* toxicity studies, the majority of them are carried out on rodent species. The choice of the route of administration depends on the physical and chemical characteristics of the test substance and the predominant route of exposure of humans. The ethnomedical information on the traditional use of the plant remedy will provide a valuable guidance for the selection of administration route in a toxicological survey. The main route of administration used is commonly oral.

For the design of repeated dose studies, the information about results of previous toxicity assays as the single dose assay data must be taken in account. The objectives of repeated dose toxicity studies (28, 90 and 180 days-exposure) include the identification of target organ(s), the characterization of the dose-response relationship, the identification of a no-observed-adverse-effect level (NOAEL), the extrapolation of carcinogenic effects to low dose human exposure levels and the prediction of chronic toxicity effects at human exposure levels (FDA, 2010).

In summary, the comprehensive assessment of the results of the studies performed must allow prediction of the toxicity potential of the evaluated substance and its possible consequences for humans.

SOME EXAMPLES OF THE RELEVANCE OF NON-CLINICAL PHARMACOLOGICAL STUDIES AS A VALIDATION OF TRADITIONAL USE OF *MANGIFERA INDICA* L. EXTRACT AND MANGIFERA PURIFIED FROM THE EXTRACT AS A MAJOR ACTIVE INGREDIENT. A WAY FOR THE DEVELOPMENT OF NEW PHYTOMEDICINAL PRODUCTS

The aqueous extract from stem bark of *Mangifera indica* L. has been used in Cuba for several years in ethnomedical practices for the improvement of quality of life of patients with different pathologies (Tamayo et al., 2001; Garrido et al., 2004; Núñez-Sellés et al., 2007a). A phytochemical characterization of the extract has led to the isolation of 9 phenolic constituents, with glucosylxanthone mangiferin as a major component, and different microelements as zinc, copper and selenium (Núñez-Sellés et al., 2002, 2007b; Romero et al., 2015).

The extract commercially identified as Vimang (Núñez-Sellés et al., 2007a), presents an adequate safety profile after all the regulatory toxicological studies (Rodeiro et al., 2006; Garrido et al., 2009; Rodeiro et al., 2012; Rodeiro et al., 2014) and a pharmacological efficacy where, according to its peculiar chemical composition, it is presented as a powerful extract with antioxidant activity, with mangiferin as a major active ingredient (Martínez et al., 2000a, b; Pardo-Andreu et al., 2005; Andreu et al., 2005a, b; Sá-Nunes et al., 2006; Rodriguez et al., 2006; Pardo-Andreu et al., 2006a, b, c, d, e, f, g; Garrido et al., 2008; Pardo-Andreu et al., 2008a, b).

Others studies have shown that the extract also possesses other pharmacological activities, such as: anti-inflammatory (Delgado et al, 2001; Garrido et al, 2004a, b; Garrido et al., 2005; Garrido et al., 2006; López-Mantecon et al., 2014); neuroprotector (Martínez et al, 2001; Lemus et al., 2009; Preissler et al., 2009; Pardo-Andreu et al., 2010; Andreu et al., 2011; Zajac et al., 2013; Maurmann et al., 2014), anti-allergic (Garcia et al, 2003a; Garcia et al., 2006), analgesic (Garrido Suarez et al., 2010; Garrido-Suarez et al., 2014a, 2014a b; Garrido-Suarez et al., 2018), hepato-protector (Remirez et al., 2005; Rodeiro et al., 2008; Rodeiro et al., 2013; Tolosa et al., 2013), antitumor by antiangiogenic mechanism (Garcia Rivera et al., 2011; Delgado-Hernandez et al., 2020) and immunomodulator (Garcia et al, 2002; Garcia et al, 2003b; Leiro et al, 2004; Hernandez et al., 2006; Hernandez et al., 2007; Garcia Rivera et al., 2011), with very complex and multifactorial mechanisms of action involved, in addition to other pharmacological activities that have been studied (Núñez-Sellés et al., 2007a, b, 2015, 2016; Garrido Suarez et al., 2020). These properties are related to its scavenger capacity of different reactive oxygen species. The interaction of mangiferin and other components of the extract with Fe2+, represents an important antioxidant mechanism characterized in different studies (Pardo-Andreu et al., 2005; Andreu et al., 2005a; Pardo-Andreu et al., 2006a, c, d).

On the other hand, mangiferin and Vimang have the property of modulating different mediators involved in immune response, more specifically: 1) inhibit nitric oxide and pro-inflammatory cytokine production in several inflammatory conditions (Garrido et al., 2004a; b) 2) inhibit phospholipase A2 (PLA2) activity and eicosanoid production (Garrido et al., 2004b; Garrido et al, 2006); 3) stimulate TGFbeta production as anti-inflammatory cytokine (Garcia et al., 2002; Garcia et al., 2011); 4) inhibit activation of transcriptional nuclear factor kB (NF-kB) (Garrido et al., 2005; Delgado-Hernandez et al., 2020); on the other hand, 5) Experiments carried out in models of apoptosis induced by the activation of T cells (activation-induced T-cell death/AICD) demonstrated a significant improvement in the survival of these cells; a finding associated with decreased oxidative stress in the activated T cell (Hernández et al., 2006; Hernández et al., 2007). These facts contribute to demonstrate the therapeutic relevance of the use of natural antioxidant supplements in the treatment of HIV/AIDS patients, where the recovery and functionality of the T cell are essential. In general, the total extract and its xanthone, mangiferin are involved in several antioxidant, anti-inflammatory, antitumoral and immunomodulatory processes, properties that confer an important therapeutic potentiality as an active ingredient in the preparation of phytopharmaceuticals products for the treatment of pathologies where oxidative stress and immunomodulator disorders are related with their etiology (Dar et al., 2005; Núñez-Sellés et al., 2007a, 2016; Delgado-Hernandez et al., 2020). Different clinical studies are conducted at this moment in order to acquire new knowledge about the therapeutic potentiality of the extract and mangiferin.

CONCLUSION

These studies went through a logical and integrated sequence of research-development, beginning with the necessary ethnomedical studies and continuing with the characterization and analytical standardization studies of isolated and purified extracts, fractions and major components. Toxicological and pharmacological investigations determine safety and pharmacological efficacy, and go deeper into the mechanisms of action characteristic of these products of natural origin, demonstrating how much can be achieved with non-clinical research to guarantee the development of controlled clinical trials. The subsequent registration of products by the regulatory entities of phytomedicines and nutraceuticals concludes with the clinical use of extracts and products that were generated from ethnomedicine and its traditional use.

ACKNOWLEDGEMENTS

We thank the revision and correction of English of the present paper made by professors Hector Luis Miguel Campos and Gilberto Suarez Balseiro from the Institute of Pharmacy and Food Sciences, University of Havana.

A special recognition to Professor Migdalia Miranda, who with her example and perseverance encouraged us to prepare this material to serve as a demonstration to the world of the importance and relevance of natural products for therapeutic purposes in our Latin American countries. Migdalia lives in every word we have written in this book and continues to be an eternal flame of light and hope for natural and traditional medicine in our Latin American peoples.

REFERENCES

Aftab, K. 2018. Natural products pharmacology. Academia Journal of Medicinal Plants 6(12): 402–403. https://doi.org/10.15413/ajmp.2018.0180

Andreu, G.P., Delgado, R., Velho, J.A., Curti, C., Vercesi, A.E. 2005a. Iron complexing activity of mangiferin, a naturally occurring glucosylxanthone, inhibits mitochondrial lipid peroxidation induced by Fe2+-citrate. Eur J Pharmacol. 513(1–2): 47–55.

Andreu, G.L., Delgado, R., Velho, J.A., Curti, C., Vercesi, A.E. 2005b. Mangiferin, a natural occurring glucosyl xanthone, increases susceptibility of rat liver mitochondria to calcium-induced permeability transition. Arch Biochem Biophys 439(2): 184–193.

Andreu, G.L., Maurmann, N., Reolon, G.K., de Farias, C.B., Delgado, R., Roesler, R. 2011. Effect of mangiferin, a naturally occurring glucoxylxanthone, on fear memory in rats. Arzneimittelforschung 61(7): 382–385.

Atanasov, A.G., Zotchev, S.B., Dirsch, V.M., Supran, C.T. 2021. Natural products in drug discovery: advances and opportunities. Nature Reviews Drug Discovery 20: 200–216. https://doi.org/10.1038/s41573-020-00114-z

Attoub, S., Arafat, K., Gelaude, A., Al Sultan, M.A., Bracke, M., Collin, P., et al. 2013. Frondoside a suppressive effects on lung cancer survival, tumor growth, angiogenesis, invasion, and metastasis. PLoS One 8:e53087.

Aydin, A., Aktayb, G., Yesilada, E. 2016. A guidance manual for the toxicity assessment of traditional herbal medicines. Natural Product Communications 11(11).

Badyal, D.K., Desai, C. 2014. Animal use in pharmacology education and research: the changing scenario. Indian Journal of Pharmacology 46(3): 257–265. https://doi.org/10.4103/0253-7613.132153

Baldim, J.L., Rosa, W., Conceição, M.F., Chagas-Paula, D.A., Ghilardi, J.H., Gomes, M. 2017. Homoisoflavonoids from Caesalpinia spp.: A closer look at chemical and biological aspects. En: Flavonoids: From biosynthesis to human health. IntechOpen. Disponible en. http://dx.doi.org/10.5772/67723

Bennett, G.J., Xie, Y.K. 1988. A peripheral mononeuropathy in rat that produces disorders of pain sensation like those seen in man. *Pain* 33: 87–107.

Bensatal, A., Rahmoun, D., Ardja, S.A., Cheikh, M., Kahouadji, A., Bekhit, M. 2020. In vitro antilithiasic activity of saponins rich fraction from the leaves of Zizyphus lotus. International Journal of Green Pharmacy 14(3): 280–285.

Bernardini, S., Tiezzi, A., Masci, V.L., Ovidi, E. 2018. Natural products for human health: an historical overview of the drug discovery approaches. Natural Product Research 32(16): 1926–1950. https://doi.org/10.1080/14786419.2017.1356838

Biswas, D. 2015. Role of reverse pharmacognosy in finding lead molecules from nature. Indian Journal of Research in Pharmacy and Biotechnology 3(4): 320–323.

Butt, J., Ishtiaq, S., Ijaz, B., Mir, Z.A., Arshad, S., Awais, S. 2018. Standardization of herbal formulations at molecular level. Journal of Pharmacognosy & Natural Products 4(1): 1–9. https://doi.org/10.4172/2472-0992.1000150

Cahlíková, L., Safratová, M., Host'álková, A., Chlebek, J., Hulcová, D., Breiterová, K., Opletal, L. 2020. Pharmacognosy and its role in the system of profile disciplines in pharmacy. Natural Product Communication 15(9): 1–7. https://doi.org/10.1177/1934578X20945450

Calixto, J.B. 2000. Efficacy, safety, quality control, marketing and regulatory guidelines for herbal medicines (phytotherapeutic agents). Brazilian Journal of Medical and Biological Research 33: 179–189.

Caridade, T.N.S., Araújo, R.D., Oliveira, A.N.A., Sousa, T.S.A., Ferreira, N.C.F., Avelar, D.S., et al. Chemical composition of four different species of the Waltheria genus. Biochemical Systematic and Ecology 2018; 80: 81–83. https://doi.org/10.1016/j.bse.2018.07.003

Chaplan, S.R., Bach, F.W., Progel, J.W., Chug, J.M., Yaksh, T.L. 1994. Quantitative assessment of allodynia in the rat paw. J. Neurosci. Methods 53: 55–63. https://doi.org/10.1016/0165/0165-0270(94)90144-9

Chen, X., Yong, C., Chen, Y. 2003. Can an *in silico* drug-target search method be used to probe potential mechanisms of medicinal plant ingredients? Natural Products Reports 20: 432–444. https://doi.org/10.1039/b303745b

Chen, H., Chunxue, B., & Xiangdong, W. 2010. The value of the lipopolysaccharide-induced acute lung injury model in respiratory medicine. *Expert Rev Respir Med* 4(6): 773–783. https://doi.org/10.1586/ers.10.71

Coderre, T.J., Xanthos, D. N., Francis, L., and Bennett, G.J. 2004. Chronic post-ischemia pain (CPIP): a novel animal model of complex regional pain syndrome-type I (CRPS-I; reflex sympathetic dystrophy) produced by prolonged hindpaw ischemia and reperfusion in the rat. Pain 112: 94–105. https://doi.org/10.1016/j.pain.2004.08.001

Coderre, T.J., Bennett, G.J. 2010. A hypothesis for the cause of complex regional pain syndrome-type I (reflex sympathetic dystrophy): pain due to deep-tissue microvascular pathology. Pain Med 11: 1224–1238. https://doi.org/10.1111/j.1526-4637.2010.00911.x

Dal Secco, D., Moreira, A.P., Freitas, A., Silva, J.S., Rossi, M.A., Ferreira, S.H., et al. 2006. Nitric oxide inhibits neutrophil migration by mechanism dependent on ICAM-1: role of soluble guanylate cyclase. Nitric Oxide 15: 77–86.

Dar, A., Faizi, S., Naqvi, S., Roome, T., Zikr-ur-Rehman, S., Ali, M., et al. 2005. Analgesic and antioxidant activity of mangiferin and its derivatives: the structure activity relationship. Biol. Pharm. Bull. 28: 596–600. https://doi.org/10.1248/bpb.28.596

Delgado, R., Garrido, G., González, D., Herrera, B., Beltrán, A., et al. 2001. Mangifera indica L extract (Vimang) as a natural antioxidant with antinociceptive and anti-inflammatory properties. Minerva Medica. 92(Suppl 1–3): 98–102.

Delgado-Hernández, R., Hernández-Balmaseda, I., Rodeiro-Guerra, I., Rodriguez Gonzalez, J.C., De Wever, O., Logie, E., et al. 2020. Anti-angiogenic effects of mangiferin and mechanism of action in metastatic melanoma. Melanoma Res 30: 39–51. https://doi.org/10.1097/CMR0000000000000647

De la Vega-Hernández, K., Antuch, M., Cuesta-Rubio, O., Núñez-Figueredo, Y., Pardo-Andreu, G.L. 2017. Discerning the antioxidant mechanism of rapanone: A naturally occurring benzoquinone with iron complexing and radical scavenging activities. J Inorg Biochem. 170: 134–147. doi: 10.1016/j.jinorgbio.2017.02.019

Disco, Y., Kumar, M., Bihari, K. 2020. Ethnomedicine for drug discovery. In: Kumar J, Shukla AC, Das G. Advances in Pharmaceutical Biotechnology: Recent Progress and Future Applications. Singapore: Springer Nature Singapore Pte Ltd. https://doi.org/10.1007/978-981-15-2195-9

Dubuisson, D., Dennis, S.G. 1977. The formalin test: a quantitative study of the analgesic effects of morphine, meperidine and brain-stem stimulation in rats and cats. Pain 4: 161–174. www.sciencedirect.com/science/article/pii/030439597901300

EMEA/CPMP/SWP. 2006. Guideline on the non-clinical investigation of the dependence potential of medicinal products. EMEA/CPMO/SWP/94227, 23 March 2006.

Ernst, E. 2005. The efficacy of herbal medicine—an overview. Fundamental & Clinical Pharmacology 19: 405–409. https://doi.org/10.1111/j.1472-8206.2005.00335.x

FDA, Guidance for Industry. 2010. M3 (R2) Nonclinical Safety Studies for the Conduct of Human Clinical Trials and Marketing Authorization for Pharmaceuticals.

FDA, Guidance for Industry. 2011. Reproductive and Developmental Toxicities Integrating Study Results to Assess Concerns.

Fidelis, Q.C., Ribeiro, T.A.N., Araújo, M.F., de Carvalho, M.G. 2014. Ouratea genus: chemical and pharmacological aspects. Brazilian Journal of Pharmacognosy 24: 1–19. https://doi.org/10.1590/0102-695X20142413361

Guide for the care and Use of Laboratory Animals as adopted and promulgated by the US National Institute of Health. NIH Publication No. 85–23, revised 1996.

García, D., Delgado, R., Ubeira, F.M., Leiro, J. 2002. Modulator efects of rat macrophage function by Mangifera indica L extract (Vimang) and mangiferin. Int Immunopharmacol. 2(6): 797–806.

García, D., Escalante, M., Delgado, R., Leiro, J. 2003a. Antihelminthic and antiallergic effect of aqueous extract of Mangifera indica l. (Vimang) ítem bark components and mangiferin. Phytother Res.17(10): 1203–1208.

García, D., Leiro, J., Delgado, R., Sanmartín, M.L., Ubeira, F.M. 2003b. Mangifera indica L Extract (Vimang) and mangiferin modulate mouse homoral immune responses. Phytother Res. 17(10): 1182–1187.

García, D., Hernández, I., Álvarez, A., Cancio, B., Márquez, L., Garrido, G., et al. 2006. Antiallergic properties of Mangifera indica L. extract (VIMANG®) and contribution of its compound mangiferin. J Pharm Pharmacol. 58(3): 382–392.

García-Rivera, D., Delgado, R., Bougarne, N., Haegeman, G., Vanden Berghe, W. 2011. Gallic acid indanone and mangiferin xanthone are strong determinants of immunosuppressive antitumour effects of Mango bark extract in MDA-MB231 breast cancer cells. *Cancer letters* 305: 21–31.

Garrido, G., Delgado, R., Lemus, Y., García, D., Beltrán, A., Rodríguez, J., Quintero, G., et al. 2004. Mangifera Indica L. (Vimang®) extract: from to ethnomedicine to clinic practice. Boletín Latinoamericano y Caribeño de Plantas Medicinales y Aromáticas (BLACPMA) 3(6): 107–109.

Garrido, G., Delgado, R., Herrera, B., Lemus, Y., García, G., Núñez-Sellés, A.J. 2004a. Protection against shock and suppression of tumor necrosis factor alpha and nitric oxide on macrophages and microglia by VIMANG, a standard aqueous extract of Mangifera indica L. Role of mangiferin isolated from the extract. Pharmacol Res. 50(2): 165–172.

Garrido, G., González, D., Lemus, Y., García, D., Lodeiro, L., Quintero, G., et al. 2004b. In vivo and in vitro anti-inflammatory activity of Mangifera indica L. extract. Pharmacol Res. 50(2): 143–149.

Garrido, G., Blanco-Molina, M., Sancho, R., Macho, A., Delgado, R., Muñoz, E. 2005. An aqueous stem bark extract of Mangifera indica (Vimang) inhibits T cell proliferation and TNF-induced activation of nuclear transcription factor NF-kappaB. Phytother Res.19(3): 211–215.

Garrido, G., Gonzalez, D., Lemus, Y., Delporte, C., Delgado, R. 2006. Protective effects of a standard extract of Mangifera indica L. (VIMANG) against mouse ear edemas and its inhibition of eicosanoid production in J774 murine macrophages. Phytomedicine 13(6): 412–418.

Garrido, G., González, D., Romay, Ch., Quintero, G., Núñez-Sellés, A.J., et al. 2008. Scavenger effect of Mangifera indica L. extract (Vimag®) on free radicals produced by human polymorphonuclear cells and hypoxanthine-xanthine oxidase chemiluminescence systems. Role of mangiferin isolated from the extract. Food Chemistry 107(3): 1008–1014.

Garrido, G., Rodeiro, I., Hernández, I., García, G., Pérez, G., Merino, N., et al. 2009. In vivo acute toxicological studies of an antioxidant extract from Mangifera indica L. (Vimang). Drug Chem Toxicol. 32(1): 53–58.

Garrido-Suárez, B.B., Garrido, G., Delgado, R., Bosch, F., Rabí, M.C. 2010. A Mangifera indica L. extract could be used to treat neuropathic pain and implication of mangiferin. Molecules 15(12): 9035–9045.

Garrido-Suarez, B.B., Garrido, G., Castro, M., Merino, N., Valdes, O., Rodeiro, I., et al. 2014a. Anti hypernociceptive effect of mangiferin in persistent and neuropathic pain models in rats. Pharmacol Biochem and Behavior 124: 311–319.

Garrido-Suarez, B.B., Castro Labrada, M., Merino, N., Valdés, O., Delgado-Hernández, R. 2014b. Anti-hyperalgesic effect of Mangifera indica L. extract on the sciatic chronic constriction injury model in rats. J Pharm Pharmacogn Res 2(2): 36–44.

Garrido-Suárez, B.B., Garrido, G., Castro-Labrada, M., Pardo-Ruíz, Z., Bellma Menéndez, A., Spencer, E., et al. 2018. Anti-allodynic Effect of Mangiferin in Rats With Chronic Post-ischemia Pain: A Model of Complex Regional Pain Syndrome Type I. Front. Pharmacol. 9: 1119. https://doi.org/10.3389/fphar.2018.01119

Garrido Suarez, B.B., Garrido, G., Piñeros, O., Delgado-Hernandez, R. 2020. Mangiferin: Possible Uses in the Prevention and Treatment of Mixed Osteoarthritic Pain. *Phytother Res* 34(3): 505–525. https://doi.org/10.1002/ptr.6546

Hacker, M. 2009. History of pharmacology-From antiquity to the twentieth century. In: Hacker M, Messer W, Bachman, K. Pharmacology. Principles and practice. California: Elsevier Academic Press. https://doi.org/10.1016/C2009-0-01474-3

Hernandez, P., Delgado, R., Walczak, H. 2006. Mangifera indica L. extract protects T cells from activation-induced cell death. Int Immunopharmacol. 6(9): 1496–1505.

Hernandez, P., Rodriguez, P.C., Delgado, R., Walczak, H. 2007. Protective effect of Mangifera indica L. polyphenols on human T lymphocytes against activation-induced cell death. Pharmacol Res. 55(2): 167–173.

ICH. 2000. International Conference on Harmonization of technical requirements for registration of pharmaceuticals for human use. Safety pharmacology studies for human pharmaceuticals, S7A

Ifeoma, O. and Oluwakanyinsola, S. 2013. Screening of Herbal Medicines for Potential Toxicities. In: New Insights into Toxicity and Drug Testing, cap 4, http://dx.doi.org/10.5772/54493

Jin, H., Hye, E., Sung, K., Sook, H., Pyo, Y. 2010. Identification of marker compounds in herbal drugs on TLC with DART-MS. Archives of Pharmacal Research 33(9): 1355–1359. https://doi.org/10.1007/s12272-010-0909-7

Kim, H.J., Jee, E.H., Ahn, K.S., Choi, H.S., Jang, Y.P. 2010. Identification of marker compounds in herbal drugs on TLC with DART-MS. Arch Pharm Res. 33(9): 1355–1359. doi: 10.1007/s12272-010-0909-7

Kumar, P., Nagarajan, A., Uchil, P.D. 2018. Analysis of Cell Viability by the MTT Assay. Cold Spring Harb Protoc. (6). https://doi.org/10.1101/pdb.prot095505
Lahlou, M. 2007. Screening of natural products for drug discovery. Expert opinion on Drug Discovery 2(5): 697–705. https://doi.org/10.1517/17460441.2.5.697
Leiro, J., Garcia, D., Arranz, J.A., Delgado, R., Sanmartin, M.L., Orallo, F. 2004. An Anacardiaceae preparation reduces the expression of inflammation-related genes in murine macrophages. Int Immunopharmacol 4(8): 991–1003.
Lemus-Molina, Y., Sánchez-Gómez, M.V., Delgado-Hernández, R., Matute, C. 2009. Mangifera indica L. extract attenuates glutamate-induced neurotoxicity on rat cortical neurons. Neurotoxicol 30(6): 1053–1058.
Li S, Han Q, Qiao C, Song J, Lung C, Xu C, Xu H. 2008. Chemical markers for the quality control of herbal medicines: an overview. Chinese Medicine 3: 7–23. https://doi.org/10.1186/1749-8546-3-7
Li, Y., Kong, D., Fu, Y., Sussman, M.R., Wu, H. 2020. The effect of developmental and environmental factors on secondary metabolites in medicinal plants. Plant Physiology and Biochemistry 148: 80–89. https://doi.org/10.1016/j.plaphy.2020.01.006
Liang, W., Chen, W., Wu, L., Li, S., Qi, Q., Cui, Y., et al. 2017. Quality evaluation and chemical markers screening of Salvia miltiorrhiza Bge. (Danshen) based on HPLC fingerprints and HPLC-MSn coupled with chemometrics. Molecules 22: 478–493. https://doi.org/10.3390/molecules22030478
Liang, H., Zhao, M., Tu, P., Jiang, Y. 2020. Polymethoxylated flavonoids from Murraya paniculata (L.) Jack. Biochemical Systematics and Ecology 93: 104–162. https://doi.org/10.1016/j.bse.2020.104162
Liu, C.X., Cheng, Y.Y., Guo, D.A., Zhang, T.J., Li, Y.Z., Hou, W.B., Huang, L.Q., Xu, H.Y. 2017. A new concept on quality markers for quality assessment and process control of Chinese medicine. Chinese Herbal Medicine 9(1): 3–13. https://doi.org/10.1016/S1674-6384(17)60070-4
López-Mantecon, A.M., Garrido, G., Delgado-Hernández, R., Garrido-Suárez, B.B. 2014. Combination of mangifera indica L. Extract supplementation plus methotrexate in rheumatoid arthritis patients: A pilot study. Phytother Res. 28(8): 1163–1172.
Louhimies, S. 2002. Directive 86/609/EEC on the protection of animals used for experimental and other scientific purposes. Altern Lab Anim. 30 Suppl 2: 217–219. https://doi.org/10.1177/026119290203002S36.
Majouli, K., Hamdi, A., Hlila, M.B. 2017. Phytochemical analysis and biological activities of Hertia cheirifolia L. roots extracts. Asian Pacific Journal of Tropical Medicine 10(12): 1134–1139. https://doi.org/10.1016/j.apjtm.2017.10.020
Manandhar, S., uitel, S., Dahal, R.K. 2019. In Vitro antimicrobial activity of some medicinal plants against human pathogenic bacteria. Journal of Tropical Medicine. https://doi.org/10.1155/2019/1895340
Martínez, G., Delgado, R., Perez, G., Garrido, G., Nunez Selles, A.J., Leon, O.S. 2000a. Evaluation of the in vitro antioxidant activity of Mangifera indica L. extract. Phytother Res. 14: 424–427.
Martínez, G., Re, L., Giuliane, A., Núñez-Sellés, A., Pérez, G., León, O.S. 2000b. Protective effects of Mangifera indica L., mangiferin and antioxidants against TPA-induced biomolecules oxidation and peritoneal macrophage activation in mice. Pharmacol Res 42(6): 565–573.
Martínez, G., Candelario-Jalil, E., Giuliane, A., León, O.S., Sam, S., Delgado, R., Núñez-Sellés, A. 2001. Mangifera indica L. Extract (Vimang) reduces ischaemia-induced neuronal loss and oxidative damage in Gerbil brain. Free Radical Res. 35: 465–473.

Mathiasen, J.R., Moser, V.C. 2018. The Irwin test and functional observational battery (FOB) for assessing the effects of compounds on behavior, physiology, and safety pharmacology in rodents. Current Protocols in Pharmacology e43. https://doi.org/10.1002/cpph.43

Maurmann, N., de Farias, C.B., Schwartsmann, G., Roesler, R., Delgado-Hernández, R., Pardo-Andreu, G.L. 2014. Mangifera indica L. extract (Vimang) improves the aversive memory in spinocerebellar ataxia type 2 transgenic mice. J Pharm Pharmacogn Res. 2(3): 63–72.

Millecamps, M., Laferriere, A., Ragavendran, J.V., Stone, L.S., and Coderre, T.J. 2010. Role of peripheral pain syndrome type 1 (CRPS-I). *Pain* 15: 174–183. https://doi.org/10.1016/j.pain.2010.07.003

Morris Ch, J. 2003. Carrageenan-induced paw edema in the rat and mouse. Methods Mol Biol 225: 115–121. https://doi.org/10.1385/1-59259-374-7:115

Mubanga, P., Mayoka, G., Mutai, P., Chibale, K. 2017. The role of natural products in drug discovery and development against neglected tropical diseases. Molecules 22: 58–99. https://doi.org/10.3390/molecules22010058

Nair, A.B., Jacob, S. 2016. A simple practice guide for dose conversion between animals and human. J Basic Clin Pharmacol 7: 27–31.

Nuñez Sellés, A.J., Vélez, H., Agüero, J., González, J., Naddeo, F., De Simone, F., et al. 2002. Isolation and quantitative analysis of phenolic compounds, polyols, fatty acids, and free sugars in an extract from mangifera indica L. (Mango) used in Cuba as nutritional supplement. J of Agricult and Food Chem 50(4): 762–766.

Núñez-Sellés, A.J., Delgado-Hernandez, R., Garrido-Garrido, G., Garcia-Rivera, D., Guevara-Garcia, M., Pardo-Andreu, G.L. 2007a. The paradox of natural products as pharmaceuticals Experimental evidences of a mango stem bark extract. Pharmacol Res. 55(5): 351–358.

Núñez-Sellés, A.J., Durruthy Rodriguez, M.D., Rodriguez Balseiro, E., Nieto Gonzalez, L., Nicolais, V., Rastrelli, L. 2007b. Comparison of major and trace element concentration in 16 varieties of Cuban mango stem bark (Mangifera indica L.). J. Agric. Food Chem. 55(6): 2176–2181 https://doi.org/org/10.1021/jf063051+

Núñez-Sellés, A.J., Villa, D.G., Rastrelli, L. 2015. Mango polyphenols and its protective effects on diseases associated to oxidative stress. Curr. Pharm. Biotechnol. 16(3): 272–280. https://doi.org/org/10.2174/1389201016031 50202143532

Núñez-Sellés, A.J., Daglia, M., Rastrelli, L. 2016. The potential role of mangiferin in cáncer treatment through its immunomodulatory, antiangiogenic, apoptotic and gene regulatory effects. Biofactors 42(5): 475–491. https://doi.org/org/10.1002/biof.1299

O'Brien, P. and Haskings JR. 2006. In vitro cytotoxicity assessment. High Content Screening: Methods in Molecular Biology 356(V): 415–425

OECD 2002a. Test No. 423: Acute Oral toxicity - Acute Toxic Class Method, OECD Guidelines for the Testing of Chemicals, Section 4, OECD Publishing, Paris. https://doi.org/10.1787/9789264071001-en

OECD 2002b. Test No. 420: Acute Oral Toxicity - Fixed Dose Procedure, OECD Guidelines for the Testing of Chemicals, Section 4, OECD Publishing, Paris. https://doi.org/10.1787/9789264070943-en

OECD 2022. Test No. 425: Acute Oral Toxicity: Up-and-Down Procedure (UDP), OECD Guidelines for the Testing of Chemicals, Section 4, OECD Publishing, Paris. https://doi.org/10.1787/9789264071049-en

OECD 2008b. Draft consensus guideline. Guidance on genotoxicity testing and data interpretation for pharmaceuticals intended for human use S2 (R1). https://www.fda.gov/media/71980/download,https://www.ema.europa.eu/en/documents/scientific-guideline/ich-guideline-s2-r1-genotoxicity-testing-data-interpretation-pharmaceuticals-intended-human-use-step_en.pdf

Okuda, K., Sakurada, Ch., Takahashi, M., Yamada, T., Sakurada, T. 2001. Characterization of nociceptive responses and spinal releases of nitric oxide metabolites and glutamate evoked by different concentrations of formalin in rats. Pain 92(1-2): 107–115. https://doi.org/10.1016/S0304-3959(00)00476-0.

Optiz, S., Hölscher, D., Oldham, N.J., Bartram, S., Schneider, B. 2002. Phenylphenalenone-related compounds: Chemotaxonomic markers of haemodoraceae from xiphidium caeruleum. J. Nat. Prod. 65(8): 1122–1130. https://doi.org/10.1021/np020083s

Pardo-Andreu, G., Delgado, R., Velho, J., Inada, N.M., Curti, C., Vercesi, A.E. 2005. Mangifera indica L (Vimang) inhibits Fe2+-citrate-induced lipoperoxidation in isolated rat liver mitochondria. Pharmacol Res. 51: 427–435.

Pardo-Andreu, G.L., Sanchez-Baldoquin, C., Avila-Gonzalez, R., Yamamoto, E.T., Revilla, A., Uyemura, S.A., et al. 2006a. Interaction of vimang (Mangifera indica L. extract) with Fe(III) improves its antioxidant and cytoprotecting activity. Pharmacol Res. 54(5): 389–395.

Pardo-Andreu, G.L., Junqueira-Dorta, D., *Delgado*, R., Cavalheiro, R.A., Santos, A.C., Vercesi, A.E., et al. 2006b. Vimang (Mangifera indica L. extract) induces permeability transition in isolated mitochondria, closely reproducing the effect of mangiferin, Vimang's main component. Chemico-Biol Interact.159: 141–148.

Pardo-Andreu, G.L., Delgado, R., Núñez-Sellés, A.J., Vercesi, A.E. 2006c. Dual mechanism of mangiferin protection against iron- induced damage to 2-deoxyribose and ascorbate oxidation. Pharmacol Res. 53(3): 253–260.

Pardo-Andreu, G.L., Cavalheiro, R.A., Dorta, D.J., Naal, Z., Delgado, R., Vercesi, A.E., et al. 2006d. Fe(III) shifts the mitochondria permeability transition-eliciting capacity of mangiferin to organelle's protection. J Pharmacol Exp Ther. 320(2): 646–653

Pardo-Andreu, G.L., Sánchez-Baldoquín, C., Ávila-González, R., *Delgado, R.*, Naal, Z., Curti, C. 2006e. Fe(III) improves antioxidant and cytoprotecting activities of mangiferin. Eur J Pharmacol. 547(1–3): 31–36.

Pardo-Andreu, G.L., Riaño Montalvo, A., Ávila González, R., *Delgado, R.* 2006f. Mangifera indica L. extract (Vimang) protects against LPS induced oxidative damage. Pharmacologyonline 3: 712–719.

Pardo-Andreu, G.L., Delgado, R., Núñez-Sellés, A.J., Vercesi, A.E. 2006g. Mangifera indica L. extract (Vimang) inhibits 2-deoxyribose damage induced by Fe (III) plus ascorbate. Phytother Res. 20: 120–124.

Pardo-Andreu, G.L., Paim, B.A., Castilho, R.F., Velho, J.A., Delgado, R., Vercesi, A.E., et al. 2008a. Mangifera indica L. extract (Vimang) and its main polyphenol mangiferin prevent mitochondrial oxidative stress in atherosclerosis-prone hypercholesterolemic mouse. Pharmacol Res. 57(5): 332–338.

Pardo-Andreu, G.L., Barrios, M.F., Curti, C., Hernández, I., Merino, N., Lemus, Y., et al. 2008b. Protective effects of Mangifera indica L extract (Vimang), and its major component mangiferin, on iron-induced oxidative damage to rat serum and liver. Pharmacol Res. 57(1): 79–86.

Pardo-Andreu, G.L., Maurmann, N., Reolon, G.K., de Farias, C.B., Schwartsmann, G., Delgado, R., et al. 2010. Mangiferin, a naturally occurring glucoxilxanthone improves long-term object recognition memory in rats. Eur. J. Pharmacol. 635(1–3): 124–128.

Petrovska, B.B. 2012. Historical review of medicinal plants' usage. Pharmacognosy Reviews 6(11): 1–5. https://doi.org/10.4103/0973-7847.95849

Phillipson, J.D. 2007. Phytochemistry and pharmacology. Phytochemistry 68: 2960–2972. https://doi.org/10.1016/j.phytochem.2007.06.028

Prasad, B. 2016. A review on drug testing in animals. Translational Biomedicine 7(4): 99–102. https://doi.org/10.2167/2172-0479.100099

Preissler, T., Martins, M.R., Pardo-Andreu, G.L., Henriques, J.A., Quevedo, J., Delgado, R., et al. 2009. Mangifera indica extract (Vimang) impairs aversive memory without affecting open field behaviour or habituation in rats. Phytother Res. 23(6): 859–862. doi: 10.1002/ptr.2713

Pugsley, M.K., Authier, S., Curtis, M.J. 2008. Review. Frontiers in Pharmacology. Principles of safety pharmacology. British J of Pharmacol 154: 1382–1399.

Remirez, D., Tafazoli, S., Delgado, R., Harandi, A.A., O'Brien, P.J. 2005. Preventing hepatocyte oxidative stress cytotoxicity with Mangifera indica L. extract (Vimang). Drug Metabol Drug Interact. 21(1): 19–29.

Qiong, Q., Zeng, L., Zhang, T., Yaron, S., Xia, H., Quan, Z., Corke, H. 2020. Phenolic profile, antioxidant and antiproliferative activities of turmeric (*Curcuma longa*). Industrial Crops and Products 152: 112561. https://doi.org/10.1016/j.indcrop.2020.112561

Rivera, A., Ortíz, O.O., Bijttebier, S., Vlietinck, A., Apers, S., Pieters, L., Catherina, C. 2017. Selection of chemical markers for the quality control of medicinal plants of the genus Cecropia. Pharmaceutical Biology 55(1): 1500–1512. https://doi.org/10.1080/13880209.2017.13074

Rodeiro, I., Cancino, L., Gonzalez, J.E., Morffi, J., Garrido, G., Gonzalez, R.M., et al. 2006. Evaluation of the genotoxic potential of Mangifera indica L. extract (Vimang), a new natural product with antioxidant activity. Food Chem Toxicol. 44(10): 1707–1713.

Rodeiro, I., Donato, M.T., Lahoz, A., Garrido, G., Delgado, R., Gómez-Lechón, M.J. 2008. Interactions of polyphenols with P450 system: possible implications on human therapeutic. Minireview in Medicinal Chem. 8(2): 97–106.

Rodeiro, I., Hernandez, S., Morffi, J., Herrera, J.A., Gómez-Lechón, M.J., Delgado, R., et al. 2012. Evaluation of genotoxicity and DNA protective effects of mangiferin, a glucosylxanthone isolated from Mangifera indica L. stem bark extract. Food Chem Toxicol. 50(9): 3360–3366.

Rodeiro, I., Gómez-Lechón, M.J., Perez, G., Hernandez, I., Herrera, J.A., Delgado, R., et al. 2013. Mangifera indica L. Extract and Mangiferin Modulate Cytochrome P450 and UDP-Glucuronosyltransferase Enzymes in Primary Cultures of Human Hepatocytes. Phytother Res. 27(5):745–752. https://doi.org/10.1002/ptr.4782

Rodeiro, I., Delgado, R., Garrido, G. 2014. Effects of Mangifera indica L stem bark extract and mangiferin on the radiation-induced DNA damage in human lymphocytes and lymphhoblastic cells. Cell Proliferation 47(1): 48–55.

Rodriguez, J., Di Pierro, D., Gioia, M., Monaco, S., Delgado, R., Coletta, M., et al. 2006. Effects of a natural extract from Mangifera indica L, and its active compound, mangiferin, on energy state and lipid peroxidation of red blood cells. Biochim Biophys Acta 1760(9): 1333–1342.

Romero, J.A., Vandama, R., López, M., Capote, M., Ferradá, C., Carballo, C., et al. 2015. Study of Physicochemical Parameters of Different Cultivars of Mangifera indica L. leaves for their Use as a Source of Mangiferin. International Journal of Pharmacognosy and Phytochemical Research. 7(3): 608–612.

Rosland, J.H., Hunskaar, S., Hole, K. 1990. Diazepam attenuates morphine antinociception test-dependently in mice. Pharmacol Toxicol. 66:382–386. http://onlinelibrary.wiley.com/doi/10.1111/j.1600-0773.1990.tb00766.x/abstract

Sá-Nunes, A., Rogério, A.P., Medeiros, A.I., Fabris, V.E., Andreu, G.P., et al. 2006. Modulation of eosinophil generation and migration by Mangifera indica L. extract (Vimang®). Int Immunopharmacol. 6(9): 1515–1523.

Sena, J.G., Xavier, H.S., Barbosa, J.M., Duringer, J.M. 2010. A chemical marker proposal for the Lantana genus: Composition of the essential oils from the leaves of Lantana radula and L. canescens. Natural Product Communications 5(4): 635–640.

Shi, Z.Q., Song, D.F., Li, R.Q., Yang, H., Qi, L.W., Xing, G.Z., et al. 2014. Identification of effective combinatorial marker for quality standardization of herbal medicines. Journal of Chromatography A 1345: 78–85. https://doi.org/10.1016/j.chroma.2014.04.015

Si, Y., Xu, X., Hu, Y., Si, H., Zhai, H. 2021. Novel quantitative structure—activity relationship model to predict activities of natural products against COVID-19. Chemical and Biological Drug Design 97: 978–983. https://doi.org/10.1111/cbdd.13822

Silva, V.G., Silva, R.O., Damasceno, S.R.B., Carvalho, N.S., Prudencio, R.S., Aragao, K.S., et al. 2013. Anti-inflammatory and antinociceptive activity of epiisopiloturine, an imidazole alkaloid from Pilocarpus microphyllus. J. Nat. Prod. 76: 1071–1077.

Suay, B., Falcó, A., Bueso, J.I., Anton, G.M., Pérez, M.T., Alemán, P.A. 2020. Tree-Based QSAR Model for Drug Repurposing in the Discovery of New Antibacterial Compounds against Escherichia coli. Pharmaceuticals 13: 431–443. https://doi.org/10.3390/ph13120431

Tamayo, D., Mari, E., González, S., Guevara, M., Garrido, G., Delgado, R., et al. 2001. Vimang as natural antioxidant supplementation in patients with malignant tumours. Minerva Medica 92(Suppl. 1–3): 95–97.

Thomford NE, Senthebane DA, Rowe A, Munro D, Seele P, Maroyi A et al. 2018. Natural Products for Drug Discovery in the 21st Century: Innovations for Novel Drug Discovery. Int. J. Mol. Sci. 19: 1578–1600. https://doi.org/10.3390/ijms190611578

Tolosa, L., Rodeiro, I., Donato, M.T., Herrera, J.A., Delgado, R., Castell, J.V., et al. 2013. Multiparametric evaluation of the cytoprotective effect of thae mangifera indica L stem bark extract and mangiferin in HepG2 cells. J. Pharm. and Pharmacol. 65:1073–1082.

US Food and Drug Administration 2005. Title 21, part 58. Good laboratory Practice for non clinical laboratories. 12: 301–303.

WHO 2019. World Health Organization. WHO global report on traditional and complementary medicine. https://creativecommons.org/licenses/by-nc-sa/3.0/igo

Xanthos, D.N., Bennett, G.J., Coderre, T.J. 2008. Norepinephrine-induced nociception and vasoconstrictor hypersensitivity in rat with chronic post-ischemia pain. Pain 137: 640–651. https://doi.org/10.1016/j.pain.2007.10.031

Xie, J., Zhang, A., Sun, H., Yan, G., Wang, X. 2018. Recent advances and effective strategies in the discovery and applications of natural products. RSC Advances 8: 812–824. https://doi.org/10.1039/c7ra09475b

Yashin, A., Yashin, Y., Xia, X., Nemzer, B. 2017. Antioxidant activity of spices and their impact on human health: A review. Antioxidants 6: 70–87. https://doi.org/10.3390/antiox6030070

Yeh, C.H., Peng, H.C., Yang, R.S., Huang, T.F. 2001. Rhodostomin, a snake venom disintegrin, inhibits angiogenesis elicited by basic fibroblast growth factor and suppresses tumor growth by a selective alpha(v)beta(3) blockade of endothelial cells. Mol. Pharmacol 59: 1333–1342.

Zajac, D., Stasinska, A., Delgado, R., Pokorski, M. 2013. Mangiferin and its traversal into the brain. Adv Exp Med Biol. 756: 105–111. doi: 10.1007/978-94-007-4549-0_14

Part II

Ecuadorian Plants with antimicrobial activity

This section introduces a description of different species of Ecuadorian plants with antimicrobial activity. The section also includes species that have been use in traditional and western medicine.

DOI: 10.1201/9781003173991-5

4 Novel List of Ecuadorian Flowers with Antimicrobial Activity

Elena Coyago-Cruz and Manuel E. Baldeón

CONTENTS

INTRODUCTION

Plants have been used by humanity as part of natural medicine, food and ornamental purposes (Arun Jyothi et al. 2011). The use of traditional medicine with natural products is common in two-thirds of countries with peripheral economies according to FAO data (FAO, FIDA, UNICEF, PMA, OMS 2018). Edible plants have chemical molecules with functional activities that can serve as dietary supplements and can be used in cosmetology (Graf et al. 2016); plants with their great variety of components have also been used to treat many human illnesses. Based on multiple ancestral knowledge, research has focused on the pursuit of new active ingredients of plant origin that can be used in the pharmaceutical, cosmetic and food industries (Villa-Ruano et al. 2013).

In recent years, flowers have received increased interest in the food industry due to their diversity in taste, smell and aesthetic value; also, flowers provide therapeutic benefits (Kaltsa et al. 2020; Fernandes et al. 2017) since they are a good source of nutrients and contain various bioactive compounds (Mlcek and Rop 2011; Kaltsa et al. 2020; Fernandes et al. 2017) such as proteins, fats, starches, amino acids, vitamins A, B, C and E, antioxidants (Lara-Cortés et al. 2013). Studies on flowers have identified the presence of different bioactive compounds such as phenolic acids, flavonoids, anthocyanins, carotenoids and antioxidants; also, studies have shown the presence of compounds with anti-inflammatory, antimicrobial, anticancer and antidiabetic activities (Lu et al. 2015). Flowers have similar nutritional composition as other parts of the plant, such as stems or leaves. Thus, in addition to their decorating properties, flowers are now being used as foods in salads, baked products, jellies, drinks, infusions and more (Coyago-Cruz et al. 2017). In this context, it is important to consider the characteristics of the flowers used as food in relation to their safety. Flowers are not traditional crops used for human consumption and could have anti-nutrients that could be potentially harmful for human health.

DOI: 10.1201/9781003173991-6

Ecuador is a mega-biodiverse country with approximately 17,000 species of vascular plants, located in different microclimates with varying temperatures and humidity (Graf et al. 2016); approximately 80% of the Ecuadorian population practices traditional medicine including the use of plants or their derived natural products (Ríos et al. 2008). However, there are limited studies on these potential useful plants from the following families: *Asteraceae, Ericaceae, Lauraceae, Myrtaceae, Myricaceae, Piperaceae, Orchidaceae, Bromeliaceae* and others (Cerón et al. 2006), *Aphelandra squarrosa, Anthurium andraeanum, Ambrosia peruviana, Dahlia pinnata, Bidens andicola, Canna indica, Senna corymbose, Salvia leucantha, Bougainvillea spectabilis, Mirabilis jalapa, Fuchsia magellanica* and *Aloysia citriodora*.

Increasing evidence indicates that plant derived products have antimicrobial, antihelminthic, antileishmanial and insecticidal activities (Hassawi et al. 2006). Plants have flavonoids, phenols, glycosides, alkaloids, carotenoids and terpenoids with antibacterial properties (Rengarajan et al. 2016). This evidence highlights the importance of the study of plants in the search for new antibiotics against microorganisms that could cause infections in animals and humans (Clark et al. 1981). Microbial pathogens are naturally found in the environment and are a frequent cause of disease in humans (Gonelimali et al. 2018). Plant derived antimicrobials present different modes of action and readily suppress infections (Longdom-Group 2021). In face of the increasing number of microbes resistant to antibiotics, there is a great need to discover new antibiotics that could be used in the clinic. This review lists updated information on specific flowers in Ecuador with antimicrobial activity and contributes new knowledge for the use of plants in health care. This information could be the basis to develop specific studies *in vitro* and *in vivo* to characterize specific components of flowers that can be applied in the pharmaceutical, food and cosmetic industries. Within this context, specific Andean species with antimicrobial activity will be listed.

***Aphelandra squarrosa* Nees:** This plant belongs to the Acanthaceae family (Figure 4.1) and the relevant synonyms are: *Aphelandra chrysops* Bull, *Aphelandra concinna* Rizzini (continues in The-Plant-List (2019)). It is native to Central and South America and the common names are: Aphelandra, Kuda Belang (Javaid et al. 2014). This species has antimicrobial activity against *Staphylococcus aureus,*

FIGURE 4.1 *Aphelandra squarrosa* flower

Escherichia coli, Proteus vulgaris, Pseudomonas aeruginosa, Salmonella typhi and *Streptococcus pneumonia* (Awan et al. 2014; Mohd et al. 2012).

Anthurium andraeanum Linden ex André: It belongs to the Araceae family (Figure 4.2) and the relevant synonyms are: *Anthurium andraeanum* var. *divergens* Sodiro (continues in The-Plant-List (2019)). It is native to Central and South America and the common names are: Anthurium, capotillo, flower of love, flamenco flower and Hawaii Emblem (JSTOR 2018; The-Plant-List 2019; Hernández 2004). This species has antimicrobial activity against *Bacillus cereus, Staphyloccocus aureus, Escherichia coli, Klebsiella pneumonia, Aspergillus fumigans* and *Penicillium chrysogenum* (Abima et al. 2016).

Ambrosia peruviana Willd: It belong to the Asteraceae family (Figure 4.3) and the relevant synonyms include *Ambrosia peruviana* DC., *Ambrosia peruviana* Cabrera and *Ambrosia peruviana* var. *Cumanensis* (Kunth) OESschulz (continues in The-Plant-List (2019)). It is native to South America and the common names are: Marco altamisa, artemisa, altamiz and alcanfor (Yánez et al. 2011; Aponte et al. 2010). This species has antimicrobial activity for *Staphylococcus aureus, Enterococcus faecalis, Escherichia coli, Salmonella typhimurium, Bacillus cereus* and *Bacillus subtilis* (Mesa et al. 2017).

Dahlia pinnata Cav.: It belongs to the Asteraceae family (Figure 4.4) and the relevant synonyms are: *Dahlia pinnata* var. *Cervantesii* (Lag. Ex Sweet) Voss and *Dahlia pinnata* var. *Coccinea* (Cav.) Voss (continues in The-Plant-List (2019)). It is native to Mexico, Central and South America and the common name is Dalia

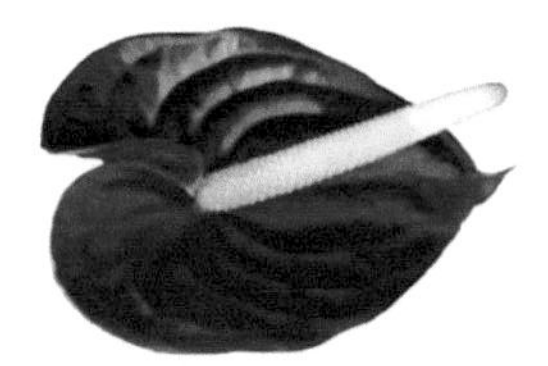

FIGURE 4.2 *Anthurium andraeanum* flower

FIGURE 4.3 *Ambrosia peruviana* flower

(Mariña 2015). This species has antimicrobial activity for *Sitophilus zeamais* and *Sitophilus oryzae* (Wang et al. 2015).

***Bidens andicola* Kunth**: It belongs to the Asteraceae family (Figure 4.5) and the relevant synonyms are: *Bidens cosmantha* Griseb and *Bidens fruticulose* Meyen & Walp., (continues in The-Plant-List (2019)). It is native to South America and the common names are Ñachac, Mìsico, Quello-ttica, Quico, Zumila (The-Plant-List 2019; López 2016). This species has antimicrobial activity against *Escherichia coli, Staphylococus aureous, Salmonella thyphimurium* and *Staphylococus enterica* (Aguilar et al. 2016; Rodríguez et al. 2011; Jerves-Andrade et al. 2014).

Canna indica L.: It belongs to the Cannaceae family (Figure 4.6) and the relevant synonyms are: *Canna achiras* Gillies ex D. Don and *Canna altensteinii* Bouché (continues in The-Plant-List (2019)). It is native to Central and South America and the common names are Achira, achera, sago, spark, Indian cane, papantla, capacho, maraca, biri, Indian shot, sagú, arawac, Imocoma, chisgua, maraca and capacho

FIGURE 4.4 *Dahlia pinnata* flower

FIGURE 4.5 *Bidens andicola* flower

FIGURE 4.6 *Canna indica* flower

(Sandoval-Herazo et al. 2018; JSTOR 2018; The-Plant-List 2019; George 2014).This species has antimicrobial activity against *Bacillus subtilis* (Al-snafi 2015).

***Senna corymbose (Lam.)* H. S. Irwin & Barneby:** It belongs to the Fabaceae family (Figure 4.7) and the relevant synonyms are: *Adipera corymbose* (Lam.) Britton & Rose and *Cassia corymbose* Lam. (continues in The-Plant-List (2019)). It is native to South America and the common names are Buttercup bush, Argentine Senna, sena del campo, rama negra and mata negra (JSTOR 2018; The-Plant-List 2019; Alcaráz et al. 2012). This species has antimicrobial activity against, *Staphylococcus aureus*, *Listeria monocytogenes*, *Escherichia coli* and *Pseudomonas aeruginosa* (Alcaráz et al. 2012).

Salvia leucantha Cav.: It belongs to the Lamiaceae family (Figure 4.8) and the relevant synonym is *Salvia leucantha* f. *iobaphes* Fernald. It is native to East Mexican and Tropical America and the common names are Mexican bush sage or sage (JSTOR 2018; The-Plant-List 2019; Pradhita et al. 2018). This species has antimicrobial activity against *Staphylococcus aureus, Bacillus subtillis, Sarcina lutea, Escherichia coli, Pseudomonas aeruginosa* and *Candida albicans* (Rajamanickam et al. 2013).

Bougainvillea spectabilis* Willd.*: It belong to the Nyctaginaceae family (Figure 4.9) and the relevant synonyms are *Bougainvillea bracteata* Pers.,

FIGURE 4.7 *Senna corumbosa* flower

FIGURE 4.8 *Salvia leucantha* flower

FIGURE 4.9 *Bougainvillea spectabilis* flower

Bougainvillea brasiliensis Raeusch, (continues in The-Plant-List (2019)). It is native to South America and the common name is Bougainvillea (Singh and Verma 2007; Juson et al. 2016; JSTOR 2018; The-Plant-List 2019; Saikia and Lama 2011). This species has antimicrobial activity against *Salmonella typhimurium, Staphylococcus aureus, Bacillus subtilis, Bacillus cereus, Enterococcus faecalis, Corynebacterium diphtheria, Streptococcus pneumonia, Klebsiella pneumonia, Escherichia coli, Enterobacter aerogenes, Pseudomonas aeruginosa* (Bharathi et al. 2016; Sidkey 2018) *Streptococcus faecalis, Micrococcus luteus, Proteus vulgaris, Serratia marcescens, Shigella flexneri, Vibrocholerae* and *Bacillus spizizenii* (Fawad et al. 2012).

Mirabilis jalapa L.: It belongs to the Nyctaginaceae family (Figure 4.10) and the relevant synonyms are *Mirabilis jalapa* var. *jalapa*, *Mirabilis jalapa* subsp. *lindheimeri* Standl (continues in The-Plant-List (2019)). It is native to South America and the common names are Don Diego at night, dompedros, parakeet, wonder of Peru and carnation (Tang et al. 2010; JSTOR 2018; The-Plant-List 2019; Gogoi et al. 2016). This species has antimicrobial activity against *Staphylococcus aureus, Salmonella typhi, Escherichia coli, Vibrio cholera* and *Bacillus subtilis* (Ullah et al. 2011).

Fuchsia magellanica Lam.: It belongs to the Onagraceae family (Figure 4.11) and the relevant synonyms are *Fuchsia araucana* F. Phil. and *Fuchsia chonotica*

FIGURE 4.10 *Mirabilis jalapa* flower

FIGURE 4.11 *Fuchsia magellanica* flower

FIGURE 4.12 *Aloysia citriodora* flower

Phil. (continues in The-Plant-List (2019)). It is native to Peru, Chile and Argentina and the common name is *Fuchsia magellanica* (Rop et al. 2012; Ruiz et al. 2015; JSTOR 2018). This species has antimicrobial activity against *Candida albicans, Bacillus subtilis, Pseudomonas aeruginosa, Staphylococcus aureus, Escherichia coli* and *Streptococcus pneumoniae* (Mølgaard et al. 2011).

Aloysia citriodora Palau: It belong to the Verbenaceae family (Figure 4.12) and the relevant synonyms are *Aloysia sleumeri* Moldenke, *Aloysia triphylla* Royle and *Verbena fragans* Salisb (continues in The-Plant-List (2019)). It is native to South America and the common names are Lemon verbena, té árabe, hierba de limón, cedrón and hierba luisa (Afrasiabian et al. 2019). This species has antimicrobial activity against *Escherichia coli, Bacillus cereus, Staphylococcus aureus* and *Pseudomonas aeruginosa* (Bahramsoltani et al. 2018; Bagher et al. 2017).

REFERENCES

Abima, J R, A Renu, M Murugan. 2016. "Phytochemical Screening and in Vitro Antimicrobial Activity of Ornamental Plant Anthurium Andraeanum." *Journal of Pharmaceutical Sciences and Research* 8 (7): 2016.

Afrasiabian, Farshid, Meharan Ardakani, Khaled Rahmani, Nammam Azadi, Zahra Alemohammad, Reza Bidaki, Mehardad Karimi, Majid Emtiazy, Mohammad Hashempur. 2019. "Aloysia Citriodora Palau (Lemon Verbena) for Insomnia Patients : A Randomized, Double – Blind, Placebo – Controlled Clinical Trial of Efficacy and Safety." *Phytotherapy Research* 33 (October 2018): 350–359. https://doi.org/10.1002/ptr.6228

Aguilar, Rosa, María Rodríguez, Eduardo Pérez. 2016. "Metabolitos Secundarios de Mezcla de Plantas Medicinales Con Acción Antibacterial Sobre Microorganismos Causantes de Infección Puerperal En La Provincia de Chachapoyas." *Pueblo Continente* 25 (2): 61–69.

Al-snafi, Ali Esmail. 2015. "Bioactive Components and Pharmacological Effects of Canna Indica-An Overview." *International Journal of Pharmacology Toxicology* 5 (2): 71–75.

Alcaráz, Lucía, Claudia Mattana, Sara Satorres, Elisa Petenatti, Marta Petenatti, Luis Del-Vitto, Analía Laciar. 2012. "Antibacterial Activity of Extracts Obtained from Senna Corymbosa and Tipuana Tipu." *Pharmacologyonline* 3 (September): 158–161.

Aponte, José, Han Yang, Abraham Vaisberg, Denis Catillo, Edith Málaga, Manuela Verástegui, Lavona Casson, et al. 2010. "Cytotoxic and Anti-Infective Sesquiterpenes Present in Plagiochila Disticha (Plagiochilaceae) and Ambrosia Peruviana (Asteraceae)." *Planta Medica* 76: 705–707.

Arun Jyothi, J., K. Venkatesh, P. Chakrapani, A. Roja Rani. 2011. "Phytochemical and Pharmacological Potential of Annona Cherimola-a Review." *International Journal of Phytomedicine* 3 (4): 439–447.

Awan, Asif, Chaudhry Ahmed, Muhammad Uzair, Muhammad Aslam, Umer Farooq, Khuram Ishfaq. 2014. "Family Acanthaceae and Genus Aphelandra : Ethnopharmacological and Phytochemical Review." *International Journal of Pharmacy and Pharmaceutical Sciences* 6 (August 2015): 1–13.

Bagher, Seyed, Amin Khaneghah, Mohamed Koubaa, Francisco Barba, Elahe Abedi, Mehrdad Niakousari, Java Tavakoli. 2017. "Extraction of Essential Oil of Aloysia Citriodora Palau Leaves Using Continuous and Pulsed Ultrasound: Kinetics, Antioxidant Activity and Antimicrobial Properties." *Process Biochemistry* 65: 197–204. https://doi.org/10.1016/j.procbio.2017.10.020

Bahramsoltani, Roodabeh, Pourouchista Rostamiasrabadi, Zahra Shahpiri, Andrés Marques, Roja Rahimi, Mohammad Farzaei. 2018. "Aloysia Citrodora Paláu (Lemon Verbena): A Review of Phytochemistry and Pharmacology." *Journal of Ethnopharmacology* 222: 34–51. https://doi.org/10.1016/j.jep.2018.04.021

Bharathi, D, P Kalaichelvan, Varsha Atmaram, S Anbu. 2016. "Biogenic Synthesis of Silver Nanoparticles from Aqueous Flower Extract of Bougainvillea Spectabilis and Their Antibacterial Activity." *Journal of Medicinal Plants Studies* 4 (5): 248–252.

Cerón, Carlos, Dorally Córdova, Carmita Reyes. 2006. "La Vegetación y Diversidad Del Bosque Nuboso Entre Sigchos y Pucayacy, Cotopaxi-Ecuador ." *Cinchonia* 7: 1–15.

Clark, Alice M., Arouk S. El-Feraly, Wen-Shyong -S Li. 1981. "Antimicrobial Activity of Phenolic Constituents of Magnolia Grandiflora L." *Journal of Pharmaceutical Sciences* 70 (8): 951–952. https://doi.org/10.1002/jps.2600700833

Coyago-Cruz, Elena, Corell Mireia, and Antonio Meléndez-Martínez. 2017. *Estudio Sobre El Contenido En Carotenoides y Compuestos Fenólicos de Tomates y Flores En El Contexto de La Alimentación Funcional.* Edited by Punto Rojo. Universida. Sevilla: Punto Rojo Libros, S.L.

FAO, FIDA, UNICEF, PMA, OMS. 2018. "El Estado de La Seguridad Alimentaria y La Nutrición En El Mundo. Fomentando La Resiliencia Climática En Aras de La Seguridad Alimentaria y La Nutrición." FAO, FIDA, UNICEF, PMA, OMS, 199. http://www.fao.org/3/a-I7695s.pdf.

Fawad, Sardar, Nauman Khalid, Waqas Asghar, Hafiz Rasul. 2012. "In Vitro Comparative Study of Bougainvillea Spectabilis ' Stand ' Leaves and Bougainvillea Variegata Leaves in Terms of Phytochemicals and Antimicrobial Activity." *Chinese Journal of Natural Medicines* 10 (6): 441–447. https://doi.org/10.1016/S1875-5364(12)60085-5

Fernandes, Luana, Susana Casal, José Alberto Pereira, Jorge A. Saraiva, Elsa Ramalhosa. 2017. "Edible Flowers: A Review of the Nutritional, Antioxidant, Antimicrobial Properties and Effects on Human Health." *Journal of Food Composition and Analysis* 60 (March): 38–50. https://doi.org/10.1016/j.jfca.2017.03.017

George, Jency. 2014. "Screening and Antimicrobial Activity of Canna Indica against Clinical Pathogens Bioactive." *International Journal for Life Sciences and Educational Research* 2 (3): 85–88.

Gogoi, Jyotchna, Khonamai Nakhuru, Rudragoud Policegoudra, Pronobesh Chattopadhyay, Ashok Rai, Vijay Veer. 2016. "Isolation and Characterization of Bioactive Components from Mirabilis Jalapa L . Radix." *Journal of Traditional and Complementary Medicine* 6: 41–47. https://doi.org/10.1016/j.jtcme.2014.11.028

Gonelimali, Faraja D, Jiheng Lin, Wenhua Miao, Jinghu Xuan, Fedrick Charles, Meiling Chen, Shaimaa R Hatab. 2018. "Antimicrobial Properties and Mechanism of Action of Some Plant Extracts against Food Pathogens and Spoilage Microorganisms." *Frontiers in Microbiology* 9 (July): 1–9. https://doi.org/10.3389/fmicb.2018.01639

Graf, Brittany, Rojas-Silva, Patricio; Baldeón, Manuel. 2016. "Discovering the Pharmacological Potential of Ecuadorian Market Plants Using a Screens-to-Nature Participatory Approach." *Journal of Biodiversity, Bioprospecting and Development* 03 (01): 1–9. https://doi.org/10.4172/2376-0214.1000156

Hassawi, Dhia, Abeer Kharma. 2006. "Antimicrobial Activity of Some Medicinal Plants against Candida Albicans." *Journal of Biological Sciences* 6 (1): 109–114.

Hernández, Loracnis. 2004. "Revisión Bibliográfica El Cultivo Del Anthurium." *Cultivo* 25 (4): 41–51.

Javaid, Asif, Muhammad Shahzad, Chaudhary Bashir, Muhammad Uzair. 2014. "In-Vitro Biological Evaluation of Crude Extract of Aerial Parts of Aphelandra Squarrosa." *Indian Research Journal of Pharmacy and Science* 1 (3): 75–79.

Jerves-Andrade, Lourdes, Fabian León-Tamariz, Eugenia Peñaherrera, Nancy Cuzco, Vladimiro Tobar, Raffaella Ansaloni, Louis Maes, Isabel Wilches. 2014. "Medicinal Plants Used in South Ecuador for Gastrointestinal Problems: An Evaluation of Their Antibacterial Potential." *Journal of Medicinal Plant Research* 8 (45): 1310–1320. https://doi.org/10.5897/JMPR2014.5656

JSTOR. 2018. "Real Jardín Botánico de Madrid." JSTOR Global Plants. https://plants.jstor.org/compilation/chlorophytum.comosum?searchUri=si%3D26%26filter%3Dnamewithsynonyms%26so%3Dps_group_by_genus_species%2Basc%26Query%3D%2528Chlorophytum%2Bcomosum%2B%2529

Juson, Albert, Maria Martinez, Johnny Ching. 2016. "Accumulation and Distribution of Heavy Metals in Leucaena Leucocephala Lam. and Bougainvillea Spectabilis Willd. Plant Systems." *Journal of Experimental Biology and Agricultural Sciences* 4 (2320): 1–6.

Kaltsa, Olga, Achillia Lakka, Spyros Grigorakis, Ioanna Karageorgou, Georgia Batra, Eleni Bozinou, Stavros Lalas, Dimitris P. Makris. 2020. "A Green Extraction Process for Polyphenols from Elderberry (Sambucus Nigra) Flowers Using Deep Eutectic Solvent and Ultrasound-Assisted Pretreatment." *Molecules* 25 (4): 1–17. https://doi.org/10.3390/molecules25040921

Lara-Cortés, Estrella, Perla Osorio-Díaz, Antonio Jiménez-Aparicio, Silvia Bautista-Baños. 2013. "Contenido Nutricional, Propiedades Funcionales y Conservación de Flores Comestibles. Revisión." *Archivos Latinoamericanos de Nutrición* 63 (April): 197–208.

Longdom-Group. 2021. "Antimicrobial Activity." Clinical Microbiology: Open Access. www.longdom.org/scholarly/antimicrobial-activity-journals-articles-ppts-list-3252.html

López, Estefanía. 2016. "Evaluación de La Actividad Antiinflamatoria y Citotoxicidad in Vitro de Bidens Andicola." *Facultad de Ciencias* Bachelor: 113. http://dspace.espoch.edu.ec/bitstream/123456789/5795/1/56T00667.pdf

Lu, Baiyi, Maiquan Li, Ran Yin. 2015. "Phytochemical Content, Health Benefits, and Toxicology of Common Edible Flowers: A Review (2000–2015)." *Critical Reviews in Food Science and Nutrition* 56 (August): S130–S148. https://doi.org/10.1080/10408398.2015.1078276

Mariña, Liudmila. 2015. "Review Cultivation of the Dahlia." *Cultivos Tropicales* 36 (1): 103–110.

Mesa, Ana, Juan Naranjo, Andrés Diez, Omar Ocampo Zulma Monsalve. 2017. "Antibacterial and Larvicidal Activity against Aedes Aegypti L . of Extracts from Ambrosia Peruviana Willd (Altamisa)." *Revista Cubana de Plantas Medicinales* 22 (1): 1–11.

Mlcek, Jiri, Otakar Rop. 2011. "Fresh Edible Flowers of Ornamental Plants—A New Source of Nutraceutical Foods." *Trends in Food Science & Technology* 22 (10): 561–569. https://doi.org/10.1016/j.tifs.2011.04.006

Mohd, Adzin, A Norhayati, Shafekh Emynur, Fazari Mohamad, M Azlina, Zahili Tajul, Zubaidi Admad. 2012. "Screening of Seven Types Terengganu Herbs for Their Potential Antibacterial Activity against Selected Food Microorganisms." *Borneo Science*, no. September: 11–27.

Mølgaard, Per, Jes Gitz, Betül Asar, Iwona Liberna, Lise Bakkestrøm, Christina Ploug, Lene Jørgensen, et al. 2011. "Antimicrobial Evaluation of Huilliche Plant Medicine Used to Treat Wounds." *Journal of Ethnopharmacology* 138: 219–227. https://doi.org/10.1016/j.jep.2011.09.006

Pradhita, Oka, Robert Manurung, Bagoes M. Inderaja, M. Yusuf Abduh, Rahma Hanifah. 2018. "Factors Affecting Biomass Growth and Production of Essential Oil from Leaf and Flower of Salvia Leucantha Cav." *Journal of Essential Oil Bearing Plants* 21 (4): 1021–1029. https://doi.org/10.1080/0972060x.2018.1506711

Rajamanickam, Manivannan, Prabakaran Kalaivanan, Ilayaraja Sivagnanam. 2013. "Antibacterial and Wound Healing Activities of Quercetin-3-O- A -L-Rhamnopyranosyl-(1–6)- β -D-Glucopyranoside Isolated from Salvia Leucantha." *International Journal of Pharmaceutical Sciences Review and Research* 22 (1): 264–268.

Rengarajan, S., V. Melanathuru, D. Munuswamy, S. Sundaram, S. T. Selvaraj. 2016. "A Comparative Study of in Vitro Antimicrobial Activity and TLC Studies of Petals of Selected Indian Medicinal Plants." *Asian Journal of Pharmaceutical and Clinical Research* 9 (5): 259–263. https://doi.org/10.22159/ajpcr.2016.v9i5.13476

Ríos, Montserrat, Rodrigo De-La-Cruz, Arturo Mora. 2008. *Conocimiento Tradicional y Plantas Útiles Del Ecuador. Saberes y Prácticas*. Quito: Abya-Yala.

Rodríguez, María, Oscar Gamarra, Fredy Pérez. 2011. "Phytochemical and Antibacterial Evaluation of Bidens Andicola HBK 'Cadillo', Alternanthera Philoxeroides (C. Mart.) Griseb. 'Lancetilla' and Celosia Sp. 'Pashquete.'" *Arnaldoa* 18 (1): 63–67.

Rop, Otakar, Jiri Mlcek, Tunde Jurikova, Jarmila Neugebauerova, Jindriska Vabkova. 2012. "Edible Flowers—A New Promising Source of Mineral Elements in Human Nutrition." *Molecules* 17 (6): 6672–6683. https://doi.org/10.3390/molecules17066684

Ruiz, Antonieta, Luis Bustamante, Carola Vergara, Dietrich Von Baer, Isidro Hermosín-gutiérrez, Luis Obando, Claudia Mardones. 2015. "Hydroxycinnamic Acids and Flavonols in Native Edible Berries of South Patagonia." *Food Chemistry* 167: 84–90. https://doi.org/10.1016/j.foodchem.2014.06.052

Saikia, H., A. Lama. 2011. "Effect of Bougainvillea Spectabilis Leaves on Serum Lipids in Albino Rats Fed with High Fat Diet." *International Journal of Pharmaceutical Sciences and Drug Research* 3 (2): 141–145. www.ijpsdr.com

Sandoval-Herazo, Luis Carlos, Alejandro Alvarado-Lassman, José Luis Marín-Muñiz, Juan Manuel Méndez-Contreras, Sergio Aurelio Zamora-Castro. 2018. "Effects of the Use of Ornamental Plants and Different Substrates in the Removal of Wastewater Pollutants through Microcosms of Constructed Wetlands." *Sustainability (Switzerland)* 10 (5). https://doi.org/10.3390/su10051594

Sidkey, Ban. 2018. "Antibacterial and Antioxidant Properties of Bougainvillea Spectabilis L. and Myrtus Communis L. Leaves Extracts." *Global Journal of Bio-Science and Biotechnology* 7 (3): 336–342.

Singh, S., Verma, A. (2007). Phytoremediation of Air Pollutants: A Review. In: Singh, S. N., Tripathi, R. D. (eds) *Environmental Bioremediation Technologies*. Springer, Berlin, Heidelberg. 293–314 https://doi.org/10.1007/978-3-540-34793-4_13

Tang, J. C., R. G. Wang, X. W. Niu, M. Wang, H. R. Chu, Q. X. Zhou. 2010. "Characterisation of the Rhizoremediation of Petroleum-Contaminated Soil: Effect of Different Influencing Factors." *Biogeosciences* 7 (12): 3961–3969. https://doi.org/10.5194/bg-7-3961-2010

The-Plant-List. 2019. "A Working List of All Plant Species." www.theplantlist.org/

Ullah, Naveed, Mir Khan, Haider Ali, Nouman Altaf, Shakoor Ahmad, Ghayour Ahmed, Minhaj Din. 2011. "Importance of White Flowered Mirabilis Jalapa with Respect to Its Phytochemical and Antimicrobial Screening." *African Journal of Pharmacy and Pharmacology* 5 (24): 2694–2697. https://doi.org/10.5897/AJPP11.437

Villa-Ruano, Nemesio, Guilibaldo G. Zurita-Vásquez, Yesenia Pacheco-Hernández, Martha G. Betancourt-Jiménez, Ramiro Cruz-Durán, Horacio Duque-Bautista. 2013. "Anti-Lipase and Antioxidant Properties of 30 Medicinal Plants Used in Oaxaca, México." *Biological Research* 46 (2): 153–160. https://doi.org/10.4067/S0716-97602013000200006

Wang, Da-cheng Cheng, Da-ren Ren Qiu, Li-na Na Shi, Hong-yu Yu Pan, Ya Wei Li, Jin Zhu Sun, Ying Jie Xue, et al. 2015. "Identification of Insecticidal Constituents of the Essential Oils of Dahlia Pinnata Cav. against Sitophilus Zeamais and Sitophilus Oryzae." *Natural Product Research* 29 (January): 1748–1751. https://doi.org/10.1080/14786419.2014.998218

Yánez, Carlos, Nurby Rios, Flor Mora, Luis Rojas, Tulia Diaz, Judith Velasco, Nahile Rios, Pablo Melendez. 2011. "Chemical Composition and Antibacterial Activity of the Essential Oil of Ambrosia Peruviana Willd from Venezuelan Plains." *Revista Peruana de Biología* 18 (August): 149–151.

5 Medicinal Plants Against *Helicobacter pylori*

Andrea Orellana-Manzano, Glenda Pilozo, Carlos Ordoñez and Patricia Manzano*

CONTENTS

HELICOBACTER PYLORI GENERALITIES AND CLINICAL MANIFESTATION

Helicobacter pylori (H. pylori) is a gram-negative rod that infects half the world's population at some point in life (Zamani et al. 2018). Its prevalence varies according to geographic region, ethnicity, race, age and socioeconomic factors (Hooi et al. 2017). The infection is causally associated with gastritis, peptic ulcer, adenocarcinoma and gastric lymphoma (Zabala et al. 2017). There may be wide variation in prevalence within the same country between urban populations with a higher economic level and rural populations (Martínez et al. 1999; Fuenmayor et al. 2002; Sepúlveda et al. 2012).

*H. pylor*i has a corkscrew-shaped spiral morphology when found in the gastric mucosa. The microorganisms adapt coccoid forms with the prolongation of the culture or when they are subjected to unfavorable growth conditions. The bacteria are 0.5 to 1.0 microns wide by 3 microns long. It has 2 to 6 monopolar flagella, which is fundamental for its mobility, and is covered by a sheath with a lipid structure, just like the outer membrane. The optimal growth temperature of *H. pylori* is 37°C, although it can develop in a range of 35 to 39°C in microaerophilia. In these conditions, its cultivation requires serum or blood supplements between 5 and 10%.

DOI: 10.1201/9781003173991-7

Once established in the gastric mucosa, this bacterium generates a chronic inflammatory process and triggers multiple clinical manifestations.

GASTRITIS: Permanent colonization of the gastroduodenal mucosa by *H. pylori* causes inflammation with a mixed infiltrate in which polymorpho-nuclear leukocytes predominate, with lymphocytes and plasma cells, which gives rise to the so-called active chronic gastritis. One of the characteristics of this infiltrate in the pediatric age is the more significant presence of lymphocytes and plasma cells and milder clinical symptoms than in adults, which is why it is called active superficial chronic gastritis. The symptoms associated with *H. pylori* gastritis are highly variable. It can be expressed as a "non-ulcer dyspepsia", characterized by pain in the epigastrium or upper hemi-abdomen, feeling of fullness, nausea and vomiting (Di Lorenzo et al. 2005).

ULCER: The association between *H. pylori* and duodenal ulcers is clear. Between 90–95% of patients present the microorganism, and the vast majority are cured by eradicating the bacteria. Concerning gastric ulcers, there is also a clear relationship. Although only 70% of this type of ulcer is associated with the presence of *H pylori*, the rest is linked to the consumption of non-steroidal anti-inflammatory drugs (Bauer and Meyer 2011).

GASTRIC CANCER: *H. pylori* infection causes superficial gastritis that causes atrophy, a precancerous condition. In 1994, the International Agency for Research on Cancer of the World Health Organization listed *H. pylori* as a carcinogenic biological agent for men (category 1) based on epidemiological evidence associating it with gastric cancer (Baker 2020; Trajkov et al. 2007). Gastric cancer is the second most common cancer globally, with 934,000 new cases per year in 2002 (8.6% of all new cancer cases) (MINSAL 2005). Its incidence varies significantly from one country to another; Chile is among the countries with the highest rates, as well as Japan, Costa Rica and Singapore (Parkin et al. 2005).

MUCOSA-ASSOCIATED LYMPHOMA TISSUE (MALT): 90% of patients with MALT lymphoma are positive for *H. pylori*. This type of lymphoma is preferably located in the antrum of the stomach since it is the area where there is lymphoid tissue. In addition, several studies support the association of *H. pylori* with this disease, since after eradication of the bacterium, regression of low-grade lymphoma has been observed (Zucca et al. 2013).

MEDICINAL PLANTS USED AGAINST *H. PYLORI*

Phyllanthus niruri

It is a member of the *Phyllanthaceae* family, commonly known as "Chanka piedra", native to the Amazon region of Ecuador, widely distributed in tropical and subtropical areas. It is used in ancestral medicine to treat asthma, arthritis, cuts, bruises, corneal opacity, conjunctivitis, flu, colds, blennorrhagia, bronchitis, constipation, colic, diabetes, dropsy, dysentery, dyspepsia, fever, gout, gonorrhea, itching, jaundice, kidney disease, anemia, leprosy, leucorrhea, malaria, menorrhagia, menstrual discomfort, obesity, proctitis, stomach pain, tenesmus, tumor, typhoid fever and vaginitis, as well as to increase appetite and stimulate the liver. This species consists of alkaloids, coumarins, flavonoids, lignins, saponins, tannins and terpenoids. It also has antioxidant, antiprotozoal, antiviral and antimicrobial activity (Kaur et al,. 2016; Chandana et al. 2021).

In its antimicrobial action, it can combat *H. pylori*, which could be due to the high content of phenolic compounds, demonstrated in *in vitro* assays where the ability of the aqueous extract of this species to inhibit the development of *H. pylori* ATCC 43579 could be related to its content of ellagitannins such as geraniin and corilagin or other non-phenolic molecules through an undiscovered mechanism; possibly these ellagitannins were partially hydrolyzed by hot aqueous extraction forming ellagic acid, gallic acid. It is also known that this mechanism acts proportionally to the dose used; on the other hand, it is not related to the blocking of oxidative phosphorylation based on proline dehydrogenase (Ranilla et al. 2012; Chandana et al. 2021). Another *in vitro* study showed the ability of the hydroalcoholic extract of this species to significantly inhibit the growth of *H. pylori* (DSMZ) 10242, which could be related to its quercetin content, which also acts as an adjuvant in the treatment of peptic ulcers. It prevents the invasion and adhesion of *H. pylori* in gastric epithelial cells, which may correspond to the effect caused by the nuclear factor (NF)-activated kappa B that leads to the release of interleukin (IL-8) in gastric epithelial cells (Kaur et al. 2016).

Ageratum conyzoides

A member of the *Asteraceae* family, commonly known as goat weed or Pedorrera, is native to Argentina, Paraguay and Brazil. *A. conyzoides* is widely distributed in tropical and subtropical areas, is used in ancestral medicine to treat constipation, fever, common wounds, burned, antimicrobial, measles, snakebite, osteoarthritis, headache and dyspnea, measles, snakebite, and osteoarthritis; it also works as anti-pneumonia, analgesic, anti-ulcer, anti-inflammatory, anti-asthma, anti-spasmodic, anti-leprosy, hemostatic, against gastrointestinal disorder, gynecological disorder and many other skin diseases (Kotta et al. 2020; Budiman and Aulifa 2020).

This species contains terpenoids, steroids, chromenes, pyrrolizidine alkaloids and flavonoids. The essential oil is composed of phenol, phenolic ester, and coumarin, which contribute to its pharmacological properties as anti-inflammatory, antinociceptive, analgesic, antipyretic, anti-diabetic, anticataleptic, antimicrobial, bronchodilator and uterine relaxant, anthelmintic, insecticide, anti-Ehrlichia and wound healing. It is also gastroprotective, anti-HIV/AIDS, toxic in cancer cells, antiulcerogenic, antitumor and anticancer, hepatoprotective, anticonvulsant, radioprotective, anticoccidial, antidote, antiprotozoal, hematopoietic and allelopathic (Kotta et al. 2020).

Its antibacterial property is attributed to its content of alkaloids, flavonoids, saponins, phenols and tannins, whose mechanism of action could be explained by the low pH of the flavonoids. Chalcone, aurone and flavone prevent *H. pylori* from adhering to gastrointestinal cells by weakening their cell wall, subsequently causing their death (Budiman and Aulifa 2020).

Impatiens balsamina L.

It is a member of the *Balsaminaceae* family, commonly known as *Balsamina.* It is native to Taiwan and is used in ancestral medicine to treat rheumatism, fractures and nail inflammation, improve circulation and soothe postpartum pain. It also has scientific reports of antifungal, antibacterial, antitumor, antipruritic and anti-anaphylactic properties. This species contains anthraquinones, naphthoquinones, flavonoids, phenolics, triterpenoids and peptides, which contribute to its antimicrobial properties against *H. Pylori* (Wang et al. 2009).

In addition, compounds have been isolated as peptides in the seeds: quinones such as 2-methoxy-1,4-naphthoquinone (MeONQ), balsaquinone, impatienol and naphthalene-1,4-dione in the petals and in the leaves, the flavonoids nicotiflorin, naringenin, quercetin, rutin, Astragalin and kaempferol. MeONQ shows anti-*H. pylori* activity in 3 ATCC strains at low concentrations by a mechanism different from traditional antibiotics, possibly exerting its action by blocking urease, cell adhesion and invasion (Wang et al. 2009; Liu et al. 2018).

An *in vitro* study showed the extract of various parts of this species to combat strains of *H. pylori* resistant to antibiotics such as ATCC (700824, 43504 and 43526) and KMUH (4917, 4952 and 4967). The acetone and ethyl acetate extract of the pod has the greatest anti-*H. pylori* effect followed by seeds, roots and stems with a MIC between 20 to 80 μg/mL. The ability of the acetone extract to decrease the survival of this bacterium is directly proportional to the dose used. The mechanism of action could be mediated by the acid pH provided by its components, which showed more significant bacterial inhibition against equal concentrations (Wang et al. 2009).

Acacia nilotica L.

It is a member of the Fabaceae family, commonly known as gum arabic. It is native to the African region, West Asia, widely distributed in tropical areas. It is used in ancestral medicine to treat infectious diseases and as an adjuvant against cancer. This species contains hydrolyzable tannins, saponins, glycosides, phenols, terpenes and flavonoids isolated from its leaves. In addition to eicosanoic acid, oleic acid, 1-hexyl-2-nitrocyclohexane, 9-octadecenal, which contribute to its antimicrobial, hyperglycemic, molluscicidal, antihypertensive, antiplatelet, demulcent, astringent, antioxidant and anti-diabetic properties (Amin et al. 2013; Sampath et al. 2020).

An *in vitro* and *silico* study showed that the aqueous extract of the aerial parts of *A. nilotica* could inhibit the growth of *H. pylori* ATCC-26695, possibly explained by the protein-ligand interaction of the bioactive compounds of the species with different protein residues (Sampath et al. 2020). Another mechanism pointed out in its anti-*H. pylori* effect is the blocking of urease, where the inhibitor as the substrate was bound to the enzyme in a non-competitive way. This hinders the colonization of the bacteria (Amin et al. 2013; Baker 2020).

Casuarina equisetifolia

A member of the Casuarinaceae family, commonly known as Australian pine, native to Australia and widely distributed in Papua New Guinea, Southeast Asia, and the Pacific Islands, is used in ancient medicine as gastroprotective, hepatoprotective and healing, and as an antiseptic. This species is constituted by carbohydrates, alkaloids, proteins, glycosides, saponins, phenolics, flavonoids, tannins, steroids, reducing sugars and triterpenoids, as well as other isolated compounds such as gallic acid, ellagic acid, epicatechin, proanthocyanidins, protocatechuic acid, p-coumaric acid, chlorogenic acid and catechin, which contribute to its antimicrobial, antibacterial, antidiabetic, cytotoxic, hypolipidemic and antiparathyroid properties, and inhibit cancer cell proliferation (Mazerand and Cock 2020; Muthuraj et al. 2019).

An *in vivo* study in albino rats showed that the ethanol extract of this species administered at 200 and 400 mg/kg orally could reduce gastric lesions. In addition, it has been reported to have an inhibitory effect on urease, which could be the way to combat strains of *H. Pylori* (Al-Snafi 2018).

Terminalia catappa L.

A member of the Combretaceae family, commonly known as Almond, is native to the Caribbean, widely distributed in tropical and subtropical areas, and is used in ancestral medicine to treat dermatitis, hepatitis, diarrhea, and pyrethritis, gastritis and urinary tract infection. Flavonoids, polyphenols and tannins constitute this species. In addition, it has been reported that the polar extracts of leaves, fruits and bark show properties such as antimicrobial, antifungal, antioxidant, antimetastatic, anti-inflammatory, hepatoprotective, mutagenic, aphrodisiac, anti-diabetic and gastroprotective (Silva et al. 2015; Ohara et al. 2020).

Through an *in vivo* study in Wistar rats, the anti-*H. pylori* activity and the anti-ulcer effect of the aqueous extract of this species were characterized. The results showed a high gastroprotective and inhibitory effect of *H. pylori*, possibly related to its content of polyphenols such as punicalagin, gallic acid, ellagic acid and punicalin, capable of attenuating oxidative stress and hypoxia-induced apoptosis; however, the underlying mechanisms of these effects have not yet been evaluated (Silva et al. 2015).

Bryophyllum pinnatum

The Crassulaceae family is commonly known as coirama, native to Madagascar, widely distributed in tropical Africa, tropical America, India, China and Australia. Coirama is used in traditional medicine to treat sores and swellings. It is also used as a hemostatic and healing herb. Alkaloids constitute this species, as well as phenols, triterpenes, glycosides, flavonoids, steroids, bufadienolides, lipids and organic acids, which contribute to its antimicrobial, antitumor, sedative, muscle relaxant, antifungal, antiulcer, antihypertensive, tocolytic, anti-diabetic, anti-inflammatory and analgesic properties (Kouitcheu et al. 2017; Fernandes et al. 2019).

The inhibitory activity of the crude extract of this species has been reported against clinical strains of *H. pylori* with a MIC of 32 μg/mL, achieving a significant reduction of the bacterial load induced in murine models (Nugraha et al. 2020; Kouitcheu et al. 2017).

Calotropis procera

It is a member of the Asclepiadaceae family, commonly known as Giant Hogweed, native to the North African region. It is used in ancient medicine to promote sexual

TABLE 5.1
Medicinal Plants for *H. pylori* Treatment and Their Mechanism

Medicinal Plant	Scientific name	Family	Origen	Plant part// extract	Study type	Experimental model	Bioactive compound	*H. pylori* effect	Mechanism	References
Balsamina/ Alegría	*Impatiens balsamina L.*	*Balsaminaceae*	Taiwán	Pods and root/ stem/leaves	Descriptive experimental study	*In vitro* with *H. pylori* bacterial strains resistant to antibiotics (ATCC 700824, ATCC 43504, ATCC 43526, KMUH 4917, KMUH 4952, KMUH 4967). The obtaining of the MICs, MBC, the time elimination test was carried out, and the effect of the environmental pH for the plant extracts was analyzed.	2-methoxy-1,4-naphthoquinone (MeONQ) and stigmasta-7,22-diene-3β-ol (spinasterol)	Bactericidal activity against CLR, MTZ, and LVX resistant *H. pylori*	The high redox potential of the hydroquinone form of MeONQ stimulates ROS production that generates apoptosis of *H. pylori*.	(Wang et al. 2011)
Acacia	*Acacia nilotica (L.)*	*Mimosaceae*	Africa, Asia Occidental	Leaves	Descriptive, experimental study	*In vitro*, phytochemical analysis using gas chromatography and mass spectroscopy protocols, identification of antagonistic activity, well diffusion assays, and MIC analysis. In silica used to validate the anticancer properties of bioactive compounds.	eicosanoic acid, oleic acid, 1-hexyl-2-nitrocyclohexane, 9-octadecenal	Antimicrobial activity by inhibition	Antimicrobial activity does not detail the mechanism.	(Sampath et al. 2020).

Pino Australiano/ Casuarina	*Casuarina equisetifolia L*	*Casuarinaceae*	Australia	Leaves	Bibliographic review (Review article)	Analytical	Gallic acid	Microbial activity	Not determined (Presumed inhibition of urease in *H. pylori*).	(Nayeem et al. 2016)
Almendro	*Terminalia catappa L.*	*Combretaceae*	Caribe	Leaves	Pre-clinical experimental study and Phytochemical Screening	Phytochemical analysis by mass spectrometry, *In vivo* with male Wistar rats weighing 180–250 g in which gastric ulcers were produced by ethanol and ischemia-reperfusion for the evaluation of the gastroprotective activity of FrAq.	Punicalagin, Punicalin, Galagic Acid	Microbial inhibition of *H. pylori* with MIC of 125.0 µg/mL.	Undetermined	(Pinheiro et al. 2015)
Chancapiedra	*Phyllanthus niruri L.*	*Phyllanthaceae*	Ecuador (Amazonian forest)	Leaves	Ethnopharmacological review and descriptive, experimental study	Analysis of the antimicrobial activity with *H. pylori*, Phytochemical Analysis with chromatography, Analysis of the antibacterial effect of the ATCC 43579 chain with Ecuadorian and Peruvian extracts dissolved in methanol.	Gallic Acid, Ellagic Acid Acid hydroxycinnamic	Damage the cell membrane of *H. pylori*	Not determined (Presumed inhibition of proteases)	(Ranilla et al. 2012; Baker 2020)
Hoja del aire	*Bryophyllum pinnatum*	*Crassulaceae*	Africa tropical and Madagascar	Leaves	Phytochemical Screening, Descriptive Experimental Study	Powdered leaves dissolved in methanol for analysis of test tube antimicrobial reaction and chromatography. Analysis of antioxidant activity.	5-methyl-4,5,7-trihydroxyl flavones and 4,3,5,7-tetrahydroxy-5-methyl-5-propenamine anthocyanidins	Microbial inhibition of *H. pylori* with MIC of 256 µg/mL	An antioxidant that protects the gastric mucosa against ROS.	(Nugraha et al. 2020)

(*Continued*)

TABLE 5.1
(Continued)

Medicinal Plant	Scientific name	Family	Origen	Plant part// extract	Study type	Experimental model	Bioactive compound	*H. pylori* effect	Mechanism	References
Algodoncillo Gigante	*Calotropis procera*	*Asclepiadaceae*	Afro-Asiatic Regions	Leaves and flowers	Bibliographic review (Review article)	Description of the antimicrobial effect of medicinal plants where the activity of *Calotropis procera* was compared and analyzed	3-O-rutinosides of quercetin, kaempferol, and isorhamnetin, besides the flavonoid 5-hydroxy-3,7-dimethoxyflavone-4′-O-β-glucopyranoside	Inhibition of urease	Inhibits *H. pylori* by stopping the progression of AGS cell growth	(Amin et al. 2013; Safavi 2015)
Hierba de chivo/ Pedorrera	*Ageratum conyzoides*	*Asteraceae*	Argentina. Paraguay, Brazil	Leaves	Comparative experimental study	*In vitro* comparative analysis of microbial activity using the disk gel dosing method, pH and viscosity evaluation	Chalcona, auron and flavone	Adhesion inhibition	The low pH of flavonoids weakens the cell wall of *H. pylori* preventing its adherence to gastrointestinal cells and causing their death.	(Budiman and Aulifa 2020; Aguirre-Mendoza 2019)
Chillangua/ Culantro Cimarrón	*Eryngium foetidium*	*Apiaceae*	Tropical America and Cameroon	Whole plant	Pre-clinical experimental study	*In vitro* MIC and MBC antimicrobial activity; and *in vivo* in Swiss albino mice with broth microdilution methodology in both tests.	caffeic acid, coumarin, benzoic acid, salicylic acid, apigenin, and p-coumaric acid	Adherence inhibition with an MCI of 64 μg/mL	By inhibiting adherence due to damage to the cell wall, it reduced the bacterial load on the gastric mucosa.	(Kouitcheu et al. 2016)

health and to treat various diseases such as leprosy, other skin disorders, tumors, piles, diseases of the spleen, abdomen, liver problems, indigestion and scarring, and also as a purgative. The leaves of this species are constituted by alkaloids and saponins, which contribute to its hypolipidemic and anticancer properties (Amin et al. 2013; Safavi et al. 2015).

An *in vitro* evaluation showed the potential of methanol, acetone and aqueous extracts of *C. procera* to combat *H. Pylori*. This is due to the ability of this species to inhibit urease by a non-competitive mechanism where the inhibitor and the substrate were bound to the enzyme in a non-competitive manner (Amin et al. 2013).

Eryngium foetidium

It is a member of the *Apiaceae* family, commonly known as C*ulantro Cimarrón or Chillangu*a, and is native to the Tropical American region. It is used in traditional medicine to treat fever, chills, vomiting, burns, hypertension, headache, earache, stomachache, asthma, arthritis, snake bites, scorpion bites, diarrhea, malaria and epilepsy. This species has a high content of calcium, iron, carotene, riboflavin, proteins and vitamins A, B and C, which contribute to its anthelmintic, anti-inflammatory, analgesic, anticonvulsant, anticlastogenic, anticancer, anti-diabetic and antibacterial properties (Kouitcheu et al. 2016; Silalahi 2021).

An *in vitro* study in cells obtained from a gastric biopsy and *in vivo* in Swiss mice demonstrated the anti-*H. pylori* capacity of *E. foetidium* methanolic extracts with a MIC of 64 µg/mL (Kouitcheu et al. 2016).

REFERENCES

Aguirre-Mendoza, Z., Jaramillo-Díaz, N., Quizhpe-Coronel, W. 2019. Arvenses asociadas a cultivos y pastizales del Ecuador. Universidad Nacional de Loja. 2018. "Arabian Medicinal Plants Possessed Gastroprotective Effects-Plant Based Review (Part 1)." *Ediloja* 1 (1): 1–216.

Al-Snafi, Ali Esmail. 2018. "Arabian Medicinal Plants Possessed Gastroprotective Effects-Plant Based Review (Part 1)." *IOSR Journal of Pharmacy* 8 (7): 77–95.

Amin, Muhammad, Farooq Anwar, Fauqia Naz, Tahir Mehmood, and Nazamid Saari. 2013. "Anti-Helicobacter Pylori and Urease Inhibition Activities of Some Traditional Medicinal Plants." *Molecules* 18 (2): 2135–2149. https://doi.org/10.3390/molecules18022135

Baker, Doha Abou. 2020. "Plants against Helicobacter Pylori to Combat Resistance: An Ethnopharmacological Review." *Biotechnology Reports* 26: e00470. https://doi.org/https://doi.org/10.1016/j.btre.2020.e00470

Bauer, Bianca, and Thomas F. Meyer. 2011. "The Human Gastric Pathogen *Helicobacter Pylori* and Its Association with Gastric Cancer and Ulcer Disease." Edited by Hajime Kuwayama. *Ulcers* 340157. https://doi.org/10.1155/2011/340157

Budiman, Arif, and Diah Lia Aulifa. 2020. "A Study Comparing Antibacterial Activity of Ageratum Conyzoides L. Extract and Piper Betle L. Extract in Gel Dosage Forms Against *Staphylococcus aureus*." *Pharmacognosy Journal* 12 (3): 473–477.

Chandana, G, R Manasa, S Vishwanath, Naik Shekhara, and M S Mahesh. 2021. "Antimicrobial Activity of *Phyllanthus niruri* (Chanka Piedra)." *IP Journal of Nutrition, Metabolism and Health Science* 3 (4):103–108. IP Innovative Publication. https://doi.org/10.18231/j.ijnmhs.2020.021

Fernandes, Júlia M., Lorena M. Cunha, Eduardo Pereira Azevedo, Estela M. G. Lourenço, Matheus F. Fernandes-Pedrosa, and Silvana M. Zucolotto. 2019. "*Kalanchoe laciniata* and *Bryophyllum pinnatum*: An Updated Review about Ethnopharmacology, Phytochemistry, Pharmacology and Toxicology." *Revista Brasileira de Farmacognosia* 29 (4): 529–558. https://doi.org/https://doi.org/10.1016/j.bjp.2019.01.012

Fuenmayor, B. A., M. E. Cavazza, H. Beltrán de Luengo, B. Gallegos, A. E. Inciarte, L. Botero, and M. Avila. 2002. "Infección Por *Helicobacter pylori* En Pacientes Con Patología Gastrointestinal Benigna." *Revista de La Sociedad Venezolana de Microbiología* 22 (1): 27–31.

Hooi, James K. Y., Wan Ying Lai, Wee Khoon Ng, Michael M. Y. Suen, Fox E. Underwood, Divine Tanyingoh, Peter Malfertheiner, et al. 2017. "Global Prevalence of Helicobacter Pylori Infection: Systematic Review and Meta-Analysis." *Gastroenterology* 153 (2): 420–429. https://doi.org/https://doi.org/10.1053/j.gastro.2017.04.022

Kaur, Baljinder, Navneet Kaur, and Vikas Gautam. 2016. "Evaluation of *Anti-Helicobacter pylori* (DSMZ 10242) Activity and Qualitative Analysis of Quercetin by HPLC in *Phyllanthus niruri* Linn." *World Journal of Pharmacy and Pharmaceutical Sciences* 5: 1691–1706.

Kotta, Jasvidianto C., Agatha B. S Lestari, Damiana S. Candrasari, and Maywan Hariono. 2020. "Medicinal Effect, *In Silico* Bioactivity Prediction, and Pharmaceutical Formulation of *Ageratum conyzoides* L.: A Review." Edited by Hans Sanderson. *Scientifica* 2020: 6420909. https://doi.org/10.1155/2020/6420909

Kouitcheu, et al., Laure. 2016. "In *Vitro* and In *Vivo* Anti-Helicobacter Activities of *Eryngium foetidum* (Apiaceae), *Bidens pilosa* (Asteraceae), and *Galinsoga ciliata* (Asteraceae) against *Helicobacter pylori*." Edited by Gail B Mahady. *BioMed Research International* 2016: 2171032. https://doi.org/10.1155/2016/2171032

Kouitcheu, Mabeku Laure Brigitte, Bertrand Eyoum Bille, Thibau Flaurant Tchouangueu, Eveline Nguepi, and Hubert Leundji. 2017. "Treatment of *Helicobacter pylori* Infected Mice with *Bryophyllum pinnatum*, a Medicinal Plant with Antioxidant and Antimicrobial Properties, Reduces Bacterial Load." *Pharmaceutical Biology* 55 (1): 603–610. https://doi.org/10.1080/13880209.2016.1266668

Liu, Qing, Xiao Meng, Ya Li, Cai-Ning Zhao, Guo-Yi Tang, Sha Li, Ren-You Gan, and Hua-Bin Li. 2018. "Natural Products for the Prevention and Management of *Helicobacter pylori* Infection." *Comprehensive Reviews in Food Science and Food Safety* 17 (4): 937–952.

Lorenzo, Carlo Di, Richard B. Colletti, Horald P. Lehmann, John T. Boyle, William T. Gerson, Jeffrey S. Hyams, Robert H. Squires Jr, Lynn S. Walker, and Pamela T. Kanda. 2005. "Chronic Abdominal Pain in Children: A Technical Report of the American Academy of Pediatrics and the North American Society for Pediatric Gastroenterology, Hepatology and Nutrition: AAP Subcommittee and NASPGHAN Committee on Chronic Abdominal Pain." *Journal of Pediatric Gastroenterology and Nutrition* 40 (3): 249–261.

Martínez, A., F Kawaguchi, J Madariaga, C. González, A. García, M. Sánchez, F. Salgado, H. Solar, C. Ortiz, and C. Andrades. 1999. "Estudio de Infección Por H Pylori En 200 Pacientes de La Octava Región." *Gastroenterol Latinoam* 10: 316.

Mazerand, Cécile, and Ian Edwin Cock. 2020. "The Therapeutic Properties of Plants Used Traditionally to Treat Gastrointestinal Disorders on Groote Eylandt, Australia." Edited by Carlos H G Martins. *Evidence-Based Complementary and Alternative Medicine* 2020: 2438491. https://doi.org/10.1155/2020/2438491

MINSAL. 2005. "Resultados I Encuesta de Salud." *Ministerio de Salud Gobierno de Chile.* https://doi.org/http://epi.minsal.cl/epi/html/invest/ENS/InformeFinalENS.pdf

Muthuraj, S., P. Muthusamy, R. Radha, and K. Ilango. 2019. "Pharmacognostical, Phytochemical and Pharmacological Review on *Casuarina equisetifolia Linn.*" *Research Journal of Science and Technology* 13 (3): 193–199. doi: 10.52711/2349-2988.2021.00029

Nayeem, N., A. Smb, H. H. Salem, and S. AHEl-Alfqy. 2016. "Gallic Acid: A Promising Lead Molecule for Drug Development." *Journal of Applied Pharmacy 8:* 1–4. https://doi.org/10.4172/1920-4159.1000213

Nugraha Ari Satia, Agka Enggar Niken Permatasari, Carina Puspita Kadarwenny, Dwi Koko Pratoko, Bawon Triatmoko, Viddy Agustian Rosyidi, Ika Norcahyanti, et al. 2020. "Phytochemical Screening and the Antimicrobial and Antioxidant Activities of Medicinal Plants of Meru Betiri National Park—Indonesia." *Journal of Herbs, Spices & Medicinal Plants* 26 (3): 303–314. https://doi.org/10.1080/10496475.2020.1734136

Ohara, Rie, Larissa Lucena Périco, Vinicius Peixoto Rodrigues, Gabriela Bueno, Ana Caroline Zanatta, Lourdes Campaner dos Santos, Wagner Vilegas, Flavia Bessi Constatino, Luis Antonio Justulin, and Clélia Akiko Hiruma-Lima. 2020. "*Terminalia catappa* L. Infusion Accelerates the Healing Process of Gastric Ischemia-Reperfusion Injury in Rats." *Journal of Ethnopharmacology* 256: 112793. https://doi.org/https://doi.org/10.1016/j.jep.2020.112793

Parkin, D Max, Freddie Bray, J Ferlay, and Paola Pisani. 2005. "Global Cancer Statistics, 2002." *CA: A Cancer Journal for Clinicians* 55 (2): 74–108.

Pinheiro Silva, Laísa, Célio Damacena de Angelis, Flavia Bonamin, Hélio Kushima, Francisco José Mininel, Lourdes Campaner dos Santos, Flavia Karina Delella, et al. 2015. "*Terminalia catappa* L.: A Medicinal Plant from the Caribbean Pharmacopeia with Anti-*Helicobacter pylori* and Antiulcer Action in Experimental Rodent Models." *Journal of Ethnopharmacology* 159: 285–295. https://doi.org/https://doi.org/10.1016/j.jep.2014.11.025

Ranilla, Lena Gálvez, Emmanouil Apostolidis, and Kalidas Shetty. 2012. "Antimicrobial Activity of an Amazon Medicinal Plant (Chancapiedra) (*Phyllanthus niruri* L.) against *Helicobacter pylori* and Lactic Acid Bacteria." *Phytotherapy Research : PTR* 26 (6): 791–799. https://doi.org/10.1002/ptr.3646

Safavi, Maliheh, Mohammadreza Shams-Ardakani, and Alireza Foroumadi. 2015. "Medicinal Plants in the Treatment of *Helicobacter pylori* Infections." *Pharmaceutical Biology* 53 (7): 939–960. https://doi.org/10.3109/13880209.2014.952837

Sampath Gattu, Douglas J. H Shyu, Neelamegam Rameshkumar, Muthukalingan Krishnan, and Nagarajan Kayalvizhi. 2020. "In *Vitro Anti-Helicobacter pylori* and Anti-Gastric Cancer Activities of *Acacia nilotica* Aqueous Leaf Extract and Its Validation Using in *Silico* Molecular Docking Approach." *Materials Today: Proceedings* 51 (4): 1675–1684. https://doi.org/https://doi.org/10.1016/j.matpr.2020.09.032

Sepúlveda, Ester, Jessica Moreno, María L Spencer, Sandra Quilodrán, Ursula Brethauer, Carlos Briceño, and Apolinaria García. 2012. "Comparación de *Helicobacter pylori* En Cavidad Oral y Mucosa Gástrica de Acuerdo a Genotipo de Virulencia (CagA y VacAm 1)." *Revista Chilena de Infectología* 29 (3): 278–283.

Silalahi, Marina. 2021. "Essential Oils and Uses of *Eryngium foetidum* L." *GSC Biological and Pharmaceutical Sciences* 15 (3): 289–294.

Trajkov, D., K. Stardelova, M. Dimitrova, J. Mishevski, and V. Serafimoski. 2007. "*Helicobacter pylori* and Gastric Carcinoma." *Prilozi* 28 (2): 39–46.

Wang, Yuan-Chuen, Deng-Chang Wu, Jyun-Ji Liao, Cheng-Hsun Wu, Wan-Yu Li, and Bi-Chuang Weng. 2009. "In *Vitro* Activity of *Impatiens balsamina* L. against Multiple Antibiotic-Resistant *Helicobacter pylori.*" *The American Journal of Chinese Medicine* 37 (04): 713–722.

Wang, Y. C., W. Y. Li, D. C. Wu, J. J. Wang, C. H. Wu, J. J. Liao, C. K. Lin. 2011. "In *Vitro* Activity of 2-methoxy-1,4-naphthoquinone and Stigmasta-7,22-diene-3β-ol from *Impatiens balsamina* L. against Multiple Antibiotic-Resistant *Helicobacter pylori*". *The Evidence-Based Complementary and Alternative Medicine* 2011. Article ID 704721, 8.

Zabala Torrres, Beatriz, Yalda Lucero, Anne J Lagomarcino, Andrea Orellana-Manzano, Sergio George, Juan P Torres, and Miguel O'Ryan. 2017. "Prevalence and Dynamics of *Helicobacter pylori* Infection during Childhood." *Helicobacter* 22 (5): e12399.

Zamani, M., F. Ebrahimtabar, V. Zamani, W. H. Miller, R. Alizadeh-Navaei, J. Shokri-Shirvani, and M. H. Derakhshan. 2018. "Systematic Review with Meta-analysis: The Worldwide Prevalence of *Helicobacter pylori* Infection." *Alimentary Pharmacology & Therapeutics* 47 (7): 868–876.

Zucca, E., Copie-Bergman, C., Ricardi, U., Thieblemont, C., Raderer, M. y Ladetto, M. 2013. "Gastric Marginal Zone Lymphoma of MALT Type: ESMO Clinical Practice Guidelines for Diagnosis, Treatment and Follow-Up." *Annals of Oncology : Official Journal of the European Society for Medical Oncology/ESMO* 24 (6): 144–148. https://doi.org/10.1093/annonc/mdt343

6 Andean Traditional Medicine Plants

The Case of "Limpia" (Cleansing Ceremony)

Adriana Orellana-Paucar

CONTENTS

DOI: 10.1201/9781003173991-8

INTRODUCTION

The Andean worldview embraces a holistic understanding of well-being. For Andean Traditional Medicine (ATM), health comprises the internal energetic equilibrium of the individual and the equivalent energetic balance with the surrounding environment. The human being should be in harmony with the family, the community, Nature and the Cosmos to achieve complete welfare (Mathez-Stiefel et al. 2007). Rupture of this energetic balance causes disease. Therefore, treatment focuses on restoring the original stability in the body and the soul (Bautista-Valarezo et al. 2020). For this purpose, plants are the primary therapeutic resources described in ATM (Houghton 1995).

The World Health Organization (WHO) promotes the rational use of complementary therapy—including herbal remedies—to ensure widespread therapeutical coverage, especially in developing countries. The application of herbal medicines as co-adjuvant of prescribed drugs could support the administration of minimum effective doses to lessen the probability of dose-dependent adverse reactions. They may contribute to improving adherence to the prescribed pharmacological therapy.

THE "LIMPIA" (CLEANSING CEREMONY) IN THE CONTEXT OF THE ANDEAN TRADITIONAL MEDICINE

The "limpia" ceremony involves medicinal plants, perfumes and holy water for energetic and spiritual cleansing. This traditional ceremony constitutes the primary treatment of psychosomatic disorders diagnosed within the context of ATM. Usually, healers complement this treatment with herbal amulets to protect individuals (Bussmann and Sharon 2009).

Within the ATM worldview, the "limpia" is a valuable method to treat diverse diseases. A common feature among them is their etiology. ATM describes their origin in the basis of body pollution caused by evil energy ("mal"). Table 6.1 depicts a summary of these Andean pathologies (Quezada et al. 1992; Cavender and Albán 2009; Armijos et al. 2014).

TABLE 6.1
Characteristics of ATM pathologies treated with the "limpia" ceremony.

Andean pathology	Cause(s)	Signs and symptoms
"Susto", "Espanto"	Frightening experiences (i.e., accidents, nightmares, an unforeseen encounter with an aggressive animal).	Anxiety, anorexia, and insomnia.
"Mal aire" (evil air)	Exposure to malevolent energy transmitted through the air of unpopulated sites (i.e., uninhabited houses, gravesites, desolate areas).	Dizziness, headache, nausea, gastralgia, syncope, and general body discomfort.
"Mal ojo"	An intense gaze directed with affection or hatred towards a person. Children are more likely to suffer from this discomfort.	Syncope, anxiety, paleness, headache, diarrhea, nausea, and fever. In children, it is common to observe asthenia, constant crying, and crusting in the eyes.

MEDICINAL PLANTS USED IN THE "LIMPIA" CEREMONY

ATM correlates the disruption of energetic balance to the presence of "mal". Medicinal plants attract "mal" from the patient's body during the "limpia" ceremony. The healer externally "cleans" the patient from head to foot using a branch of scented plants tied as a broom. Healers use plants exhibiting a strong aroma based on the belief that the intense perfume of the flowers and leaves draws "mal" from inside the patient's body to the broom of medicinal plants. Healers complement the treatment by asking the patients to inhale the branches of the plants (Cavender and Albán 2009; Armijos et al. 2014).

The ATM wisdom assumes that there is no complete extinction of "mal". For this reason, as soon as the "limpia" ceremony finishes, the healer cautiously disposes of the plants in an uninhabited place allowing evil energy ("mal") to return gradually to Nature. "Mal" will be neutralized by "Pachamama" (Mother Earth) (Cavender and Albán 2009). The Andean worldview considers medicinal plants a virtual channel to convey Pachamama's healing energy. Figure 6.1 portrays an ATM woman healer performing the "limpia" ceremony.

The main characteristics of the medicinal plants used in the "limpia" ceremony are described below. Emphasis is pointed towards bioactive compounds present in these plants. In the context of the "limpia" ceremony, the phytochemistry and the biological activities focus on essential oils and those compounds administered topically. Regarding traditional uses, other applications, besides the "limpia", are mentioned.

ALISO

Botanical name: *Alnus acuminata* subsp. acuminata
Other local names: aliso, rambrán, ranrán

FIGURE 6.1 The "limpia" ceremony uses medicinal plants and wooden daggers (amulets) to protect the patient against "evil". (Author: Matias Villacís-Luzuriaga. Cuenca, Ecuador.)

FIGURE 6.2 *Alnus acuminata*. (Author: Matias Villacís-Luzuriaga. Cuenca, Ecuador.)

BOTANICAL INFORMATION AND GEOGRAPHICAL DISTRIBUTION

Family: Betulaceae. *Alnus acuminata* is a tree found in the colder sites of the Northern Hemisphere and at high altitudes in North and South America. The trees can reach around 30 meters in height (Russo 1990). Figure 6.2 shows an *A. acuminata* botanical specimen.

TRADITIONAL MEDICINE USES

The bark of young branches and leaves are commonly used for therapeutic purposes (Molina Vélez 2008). The leaves are rubbed against the skin to treat rheumatism. For leg and knee pain, a poultice with the leaves is applied. An infusion prepared with the whole plant is orally administered for fever (Ríos et al. 2007).

PHYTOCHEMISTRY OF PLANT

Tannins, flobaphenes, triterpenoids, diarylheptanoids, emodin and alnulin have been isolated from the bark (Molina Vélez 2008; Aguilar et al. 2011). In addition, δ-amirone and 4′,7-dimethoxy apigenin were identified in the leaves (Salama and Avendaño 2021).

BIOLOGICAL ACTIVITIES

δ-amirone and 4′,7-dimethoxy apigenin isolated from the leaves possess anti-inflammatory properties (Salama and Avendaño 2021). Likewise, the triterpenoids and diarylheptanoids extracted from the stem bark have shown anti-inflammatory activity (Aguilar et al. 2011).

Altamisa

Botanical name: *Ambrosia arborescens* Mill.
Other local name: marco

BOTANICAL INFORMATION AND GEOGRAPHICAL DISTRIBUTION

Family: Asteraceae. It is a shrub of 1.5 to 3 meters in height, typically found in riverbanks located at 2,000–3,000 meters of altitude (Cano de Terrones 2014). Figure 6.3 displays an *A. arborescens* botanical specimen.

TRADITIONAL MEDICINE USES

The whole plant is employed in traditional medicine. This plant is used to prepare a bath for the puerperium stage in women (Quezada et al. 1992).

PHYTOCHEMISTRY OF PLANT

Six new terpene derivatives were identified in the leaves: eudesm-11(13)-en-4b,9b-diol, 15R,16-dihydroxy-3-oxoisopimar-9(11)-ene, 15S,16-dihydroxy-3-oxoisopimar-9(11)-ene, 1a-hydroxy-7-oxo-iso-anhydrooplopanone, 10a-hydroxy-11,13-dihydro-5-epi-psilostachyin and 4b-hydroxypseudoguaian-12,6-olide 4-O-b-D-glucopyranoside. In addition, 12 known sesquiterpenes were also isolated from the leaves: damsin, psilostachyin, volenol, 12- hydroxy-4(5),11(13)-eudesmadien-15-al, dihydrocoronopilin, coronopilin, 4(15)-eudesmene-1b,7a-diol, 13-hydroxy-4-oxo-7(11)-

FIGURE 6.3 *Ambrosia arborescens*. (Author: Matias Villacís-Luzuriaga. Cuenca, Ecuador.)

pseudoguaien-12,6-olide, 4-(3-oxobutyl)-5-methyl-1-acetic acid-a-methylene3-cycloheptene, damsinic acid, psilostachyin C and 10a-hydroxydamsin (de Leo et al. 2010). Moreover, a specific HPLC-DAD method to further characterize the sesquiterpene lactones, coronopilin and damsi, isolated from the aerial parts of the plants, has been described (Svensson et al. 2018).

BIOLOGICAL ACTIVITIES

Coronopilin and damsin have shown anti-inflammatory activity *in vitro* by inhibiting pro-inflammatory IL-6 and MCP-1 expression in human skin cells (Svensson et al. 2018).

Anti-cancer activity has been reported to damsin and ambrosin *in vitro*. Both inhibited cell migration and cancer stem cell proliferation (Villagomez et al. 2013; Sotillo et al. 2017).

Chilca

Botanical name: *Baccharis latifolia* (Ruiz & Pav.) Pers.
Other local name: chilco

BOTANICAL INFORMATION AND GEOGRAPHICAL DISTRIBUTION

Family: Asteraceae. *Baccharis latifolia* grows spontaneously on riverbanks at 2,000–4,000 meters of altitude. It is a shrub reaching up to 2 meters in height (Sequeda-Castañeda et al. 2015). Figure 6.4 presents a *B. latifolia* botanical specimen.

TRADITIONAL MEDICINE USES

A poultice made from the leaves is applied topically to alleviate muscle pain (Ríos et al. 2007). Other traditional applications include flatulence, diarrhea, inflammation, stomachache and insomnia (Sequeda-Castañeda et al. 2015).

PHYTOCHEMISTRY OF PLANT

The leaves contain resin, oxidases, baccharin, turpetine, gallic tannins, quercetin, rutin, germacrene derivatives, diterpenes, clerodanos, labdanos, eudesmane, limonene, β-cubebeno, germacrene, δ-cadinene, epi-α-bisabolol, dihydroisochromolaenin, ledol, friedelin and dimethoxyflavone (Sequeda-Castañeda et al. 2015).

Further characterization of the essential oil extracted by hydro-distillation from the aerial parts involved gas chromatography/mass spectrometry (GC/MS) and GC/flame ionization detector (GC/FID) analyses (Valarezo et al. 2013). Its major

FIGURE 6.4 *Baccharis latifolia*. (Author: Matias Villacís-Luzuriaga. Cuenca, Ecuador.)

constituents are monoterpenes (α-pinene, limonene, α-thujene), sesquiterpenes (caryophyllene, bicyclo germacrene) and oxygenated sesquiterpene (spathulenol) (Loayza et al. 1995; Pellegrini et al. 2017).

BIOLOGICAL ACTIVITIES

Baccharis latifolia has shown anti-inflammatory activity via enzymatic inhibition of arachidonate cascade (Abad et al. 2006). The essential oil presented antibacterial activity against *Staphylococcus aureus* (Sequeda-Castañeda et al. 2015) and anti-fungal properties against *Trichophyton rubrum*, *Trichophyton mentagrophytes* and *Aspergillus fumigatus* (Zapata et al. 2010; Valarezo et al. 2013).

CHILCHIL

Botanical name: *Tagetes terniflora* Kunth
Other local names: chilchi, chilchi huandura, hierba de gallinazo

BOTANICAL INFORMATION AND GEOGRAPHICAL DISTRIBUTION

Family: Asteraceae. *Tagetes terniflora* is an herb native from South America. It grows between 2,000–3,500 meters of altitude (Zapata-Maldonado et al. 2015). Figure 6.5 exhibits a *T. terniflora* botanical specimen.

FIGURE 6.5 *Tagetes terniflora*. (Author: Matias Villacís-Luzuriaga. Cuenca, Ecuador.)

TRADITIONAL MEDICINE USES

Leaves are used for bath. Branches are used against insects (Ríos et al. 2007).

PHYTOCHEMISTRY OF PLANT

The main constituents of the essential oil were identified through gas chromatography/mass spectrometry (GC/MS): valeric acid, trans-tagetone and trans-ocimene (Zapata-Maldonado et al. 2015).

BIOLOGICAL ACTIVITIES

The essential oil showed a modest antifungal activity on *Fusarium* (Galvez et al. 2018, 2–3). Ocimene possessed pesticidal activity against *Metopolophium dirhodum* (Walk.) (Sánchez-Chopa and Descamps 2012).

EUCALIPTO

Botanical name: *Eucalyptus globulus* Labill.
Other local name: eucalipto blanco

BOTANICAL INFORMATION AND GEOGRAPHICAL DISTRIBUTION

Family: Myrtaceae. This tree is native to Australia and is distributed worldwide. It reaches up to 30–40 meters in height and proliferates in humid environments (Molina Vélez 2008). Figure 6.6 shows an *E. globulus* botanical specimen.

FIGURE 6.6 *Eucalyptus globulus*. (Author: Matias Villacís-Luzuriaga. Cuenca, Ecuador.)

TRADITIONAL MEDICINE USES

An infusion prepared with the leaves and stems is drunk to treat knee and leg pain and cough. For sleeping well or treating flu, the essential oils from the leaves are inhaled (Ríos et al. 2007).

PHYTOCHEMISTRY OF PLANT

Among the main constituents of the essential oil extracted from the leaves are oxygenated monoterpenes, monoterpenes and oxygenated sesquiterpenes. Of these, the main oxygenated monoterpene is 1,8-cineole (70%), the main sesquiterpene is α-pinene, and the main sesquiterpene is (-)-globulol (2.77%) (Dhakad et al. 2018).

BIOLOGICAL ACTIVITIES

The hexane extract from the leaves of *E. globulus* inhibited IgE-dependent histamine release from RBL-2H3 cells (Ikawati et al. 2001). In addition, the main constituent of the essential oil, 1,8-cineole, is a potent cytokines suppressor *in vitro* (Juergens et al. 1998). Also, 1,8-cineole exhibited wound healing activity in mice (Tomen et al. 2017).

The aqueous extract of *E. globulus* decreased blood glucose in diabetic rats and reduced reactive oxygen species (ROS), a pathogenic factor in type 2 diabetes. ROS attenuation is associated with the potent antioxidant activity of compounds present in the extract (Dey and Mitra 2013).

E. globulus oil from the leaves presented anthelmintic activity against the Indian earthworm *Pheretima posthuma* due to the presence of borneol, linalool, cineole, geranyl acetate, anethol and saffrol (Taur et al. 2010).

The essential oil showed antimalarial activity against *Plasmodium falciparum in vitro* (Mllhau et al. 1997). Eucalyptone G, a phloroglucinol derivative present in the essential oil of *E. globulus*, presented antibacterial activity against *Bacillus subtilis*, *Staphylococcus aureus* (Mohamed and Ibrahim 2007). The action against *Escherichia coli* is probably associated with 1,8-cineole (Bachheti et al. 2003). Also, 1,8-cineole possessed activity against respiratory bacteria (*Mycobacterium tuberculosis)*, viruses (*Epstein Barr, influenza A (H1N1)*) and fungus (*Candida*). These properties, together with the immune-stimulant, anti-inflammatory, antioxidant, analgesic and spasmolytic activities reported for the *E. globulus* oil and the safety of its usage via inhalation, support its application for respiratory problems (Sadlon and Lamson 2010; Brochot et al. 2017). In-silico studies suggested 1,8-cineole as a potential candidate to treat SARS-CoV-2 infections via inhibition of the Mpro/3CLpro proteinase, crucial for coronavirus reproduction (Sharma and Kaur 2020).

The crude extracts of fruits and leaves exerted antioxidant properties. This activity seems higher in the fruit than in the leaf extract (Said et al. 2016).

Inhalation of *E. globulus* oil decreased pain and blood pressure in patients with total knee replacement (Jun et al. 2013).

GUANDO

Botanical name: *Brugmansia arborea* L. Lagerh.; *Brugmansia sanguinea* (Ruiz & Pav.) D. Don
Other local names: Floripondio, guanto, huanduj, yurac huantuc, guantug, guando blanco, wantuc

BOTANICAL INFORMATION AND GEOGRAPHICAL DISTRIBUTION

Family: Solanaceae. *Brugmansia arborea* and *Brugmansia sanguinea* are native from America. They grow at 1,000–4,000 meters of altitude. This tree is commonly known as the angel's trumpet due to the shape of its flowers. The flowers of *B. sanguinea* are red, and the flowers of *B. arborea* vary from white to ivory color. Both possess a strong aroma (Molina Vélez 2008). Figure 6.7 displays *B. arborea* and *B. sanguinea* botanical specimens.

TRADITIONAL MEDICINE USES

The plant is used mainly for ornamental purposes. ATM uses it as a hallucinogenic beverage for rituals and ceremonies. A poultice of the plant is employed to cure stomach cramps and colic. The liquid of the mashed stem is orally administered against parasites. The flowers are used to prepare a bath to treat "espanto" (Ríos et al. 2007).

PHYTOCHEMISTRY OF PLANT

B. arborea mainly contains tropane alkaloids: atropine and scopolamine. NMR, IR, UV and MS analysis allowed the identification of 5 flavonoids in *B. arborea*

FIGURE 6.7 *Brugmansia arborea* and *Brugmansia sanguinea*. (Author: Matias Villacís-Luzuriaga. Cuenca, Ecuador.)

flowers: kaempferin, kaempferitrin, kaempferol 3-O-β-Dglucopyranosyl- 7-O-α-L-rhamnopyranoside, quercetin 3,7-di-O-α-L-rhamnopyranoside and (+)- aromadendrin (Pino and Alvis 2009; Kim et al. 2020).

BIOLOGICAL ACTIVITIES

The essential oil of *B. arborea* displayed insecticidal activity against *Haematobia irritans* (Cruz-Carrillo et al. 2011). Its anticholinergic activity is associated with its well-known alkaloids. The flavonoids of the flowers showed anti-inflammatory activity (Kim et al. 2020).

Higüila

Botanical name: *Monnina salicifolia* Ruiz & Pav.
Other local names: Azulina, higuillan, higüilán, huigra

BOTANICAL INFORMATION AND GEOGRAPHICAL DISTRIBUTION

Family: Polygalaceae. *Monnina salicifolia* is native to the center and south of Ecuadorian Andes and Northern Peru. It is a shrub growing spontaneously at the riverbanks, reaching up to 2–5 meters in height. It develops at 2,000–3,5000 meters of altitude. It has a yellow apex, and its corolla is made up of 5 blue or lilac petals (Minga 2000). Figure 6.8 presents a *Monnina salicifolia* botanical specimen.

FIGURE 6.8 *Monnina salicifolia*. (Author: Matias Villacís-Luzuriaga. Cuenca, Ecuador.)

TRADITIONAL MEDICINE USES

The flowers are used as a pigment. The leaves are utilized as food for cattle and guinea pigs. The crushed roots are employed to wash the hair and prevent dandruff (Minga 2000; Ríos et al. 2007).

PHYTOCHEMISTRY OF PLANT

No phytochemistry information has been reported to date for this plant.

BIOLOGICAL ACTIVITIES

No biological activity has been registered to date for this plant.

MORTIÑO

Botanical name: *Solanum nigrescens* M. Martens & Galeotti
Other local names: Hierba mora

BOTANICAL INFORMATION AND GEOGRAPHICAL DISTRIBUTION

Family: Solanaceae. It is a native herb that grow up to 1.5 m tall. When maturity is complete, the flowers have white or purple corolla and black fruits (Molina Vélez 2008). Figure 6.9 exhibits *S. nigrescens* botanical specimen.

FIGURE 6.9 *Solanum nigrescens*. (Author: Matias Villacís-Luzuriaga. Cuenca, Ecuador.)

TRADITIONAL MEDICINE USES

It is used to treat wounds and skin infections such as acne, abscesses and ulcers (Cárdenas et al. 2017).

PHYTOCHEMISTRY OF PLANT

The plant contains alkaloids, tannins, saponins and free anthracene derivatives (Peralta Gómez et al. 2013).

BIOLOGICAL ACTIVITIES

S. nigrescens displayed activity against *Candida albicans* when topically applied (Girón et al. 1988). Antibacterial activity against *Staphylococcus aureus* and *Streptococcus pyogenes* is also reported (Caceres et al. 1991).

PALO SANTO

Botanical name: *Bursera graveolens* (Kunth)
Other local name: guayaco

BOTANICAL INFORMATION AND GEOGRAPHICAL DISTRIBUTION

Family: Burseraceae. Palo santo is a South American native plant, growing naturally near the sea on Ecuadorian and Peruvian coasts. It typically develops at 25–1,300 meters of altitude (Molina Vélez 2008). Figure 6.10 shows a *B. graveolens* botanical specimen.

FIGURE 6.10 *Bursera graveolens*. (Author: Matias Villacís-Luzuriaga. Cuenca, Ecuador.)

TRADITIONAL MEDICINE USES

The bark is burned to exert its insecticidal activity (Tene et al. 2007). The dried leaves of *B. graveolens* prepared as an oral infusion are administered to treat dermatitis and rheumatism. It also possesses analgesic and diaphoretic properties (Fon-Fay et al. 2019). It is also employed to treat stomachache and labor complications during childbirth (Sánchez-Recillas et al. 2020).

PHYTOCHEMISTRY OF PLANT

GC-MS analysis of the essential oil obtained from the aerial parts (3.4% yield) identified limonene as the principal constituent (Monzote et al. 2021; Young 2007). GC-MS analysis of *B. graveolens*, collected on the coast of Ecuador, reported the sesquiterpene viridiflorol as the main compound present in the essential oil extracted from dried branches (Manzano Santana et al. 2009).

BIOLOGICAL ACTIVITIES

The *B. graveolens* oil possesses a mild scavenging effect (Fon-Fay et al. 2019). In addition, the anti-proliferative activity of the essential oil suggests it is a potential candidate for cancer treatment (Monzote et al. 2021).

The hexane and dichloromethane extracts from the leaves showed *in vitro* antibacterial activity against *Escherichia coli, Salmonella enterica* and *Salmonella choleraesuis.* Also, these extracts exerted spasmolytic activity in isolated rat ileum (Sánchez-Recillas et al. 2020).

Anti-inflammatory properties have been reported for the essential oil extracted from the dried branches of the plant (Manzano Santana et al. 2009).

Also, the activity against *Tribolium castaneum* cereal crop pest has been attributed to this essential oil (Fernández-Ruiz et al. 2018).

POLEO

Botanical name: *Minthostachys mollis* (Kunth) Griseb.
Other local names: yurac tipo, tipo blanco, poleo blanco del cerro

BOTANICAL INFORMATION AND GEOGRAPHICAL DISTRIBUTION

Family: Lamiaceae. This plant is native to South America. It is an aromatic herb with characteristic white flowers. It grows up to 0.5 m in height in humid places (Molina Vélez 2008). Figure 6.11 displays a *M. mollis* botanical specimen.

TRADITIONAL MEDICINE USES

The leaves are rubbed against the skin to treat the flu (Ríos et al. 2007).

FIGURE 6.11 *Minthostachys mollis*. (Author: Matias Villacís-Luzuriaga. Cuenca, Ecuador.)

PHYTOCHEMISTRY OF PLANT

The major constituents of the essential oil of *M. mollis* are neomenthol, menthol, menthone, piperitone and carvacryl acetate (Schmidt-Lebuhn 2008).

BIOLOGICAL ACTIVITIES

The essential oil of *M. mollis* exhibited antibacterial properties against *Bacillus subtilis* and *Salmonella typhi* (Mora et al. 2009).

Romero

Botanical name: *Rosmarinus officinalis* L.
Other local name: romero del cerro

BOTANICAL INFORMATION AND GEOGRAPHICAL DISTRIBUTION

Family: Lamiaceae. This plant is native to the Mediterranean region and distributed worldwide. It is a perennial shrub reaching about 2 meters in height. The aromatic green leaves fully cover the branches. It is commonly used as a spice for cooking, ornament and traditional medicinal applications (de Oliveira et al. 2019). Figure 6.12 presents a *R. officinalis* botanical specimen.

FIGURE 6.12 *Rosmarinus officinalis*. (Author: Matias Villacís-Luzuriaga. Cuenca, Ecuador.)

TRADITIONAL MEDICINE USES

The infusion of the twigs is topically applied for eye irritation, vaginal washes and skin rash. The sap from crushed leaves is directly spread over the external ear canal to treat pain (Ríos et al. 2007).

PHYTOCHEMISTRY OF PLANT

Terpene compounds were the predominant compound class identified by GC-MS analysis: 1,8-cineole (50.63%), camphor (13.27%) and α -pinene (10.11%) (Feriotto et al. 2018).

BIOLOGICAL ACTIVITIES

Rosemary oil showed free radical scavenging capacity and hepatoprotective activity (Rašković et al. 2014). The essential oil promoted wound healing in rats by topical application. This action appears to be associated with 1,8-cineole and α -pinene (Labib et al. 2019).

The essential oil exhibited anti-inflammatory activity by inhibition of arachidonic acid cascade and NF-κB transcription. Also, the *R. officinalis* oil generated smooth muscle relaxation. The antioxidant capacity of the oil is related to the activity of β-pinene, limonene, γ-terpinene, linalool, terpin-4-ol, α-terpineol and β-caryophyllene for oxygen radical absorption (Borges et al. 2019). Moreover, eucalyptol (1,8-cineole) showed *in vitro* pro-apoptotic activity by regulating the p53 signaling pathway. Hence, it is a potential candidate for carcinoma treatment (Sampath et al. 2018).

Camphor is an immunomodulator capable of increasing the phagocytic response in mice (Lin et al. 2018). There is evidence of a remarkable selection capacity of camphor carboxylate complexes for ovarian cancer cells (Carvalho et al. 2018). Camphor also presented antifungal activity against *Candida albicans* (Manoharan et al. 2017). The antimicrobial activity of the oil and its main constituents was evaluated against bacteria (*Staphylococcus epidermidis, Staphylococcus aureus* and *Bacillus subtilis*) and fungi (*Candida albicans* and *Aspergillus niger*). The rosemary oil exhibited higher antibacterial and antifungal activity than the isolated main compounds (1,8-cineole and α -pinene) (Jiang et al. 2011).

Inhalation of rosemary oil improved stress-related psychiatric disorders by decreasing serum corticosterone and increasing brain dopamine concentrations in mice. These pharmacological effects are probably associated with the anxiolytic properties of α-pinene (Villareal et al. 2017). Other medical applications on nervous system disorders include memory boosting and Alzheimer's treatment (Benny and Thomas 2019), (Rahbardar and Hosseinzadeh 2020).

RUDA

Botanical name: *Ruta graveolens* L.
Other local name: salballal

BOTANICAL INFORMATION AND GEOGRAPHICAL DISTRIBUTION

Family: Rutaceae. It is a native herb to Europe and distributed worldwide. It grows even in hostile conditions. This plant is used as an ornament and for medicinal purposes (Molina Vélez 2008). Figure 6.13 exhibits a *R. graveolens* botanical specimen.

TRADITIONAL MEDICINE USES

An infusion of this plant is drunk to expel bad energy (evil spirits) from the body. The infusion of the leaves is an emmenagogue. This infusion is also used to treat pneumonia. To treat colic, an infusion prepared with the seeds is drunk. For fever, the leaves macerated in water are topically applied on the forehead (Ríos et al. 2007).

FIGURE 6.13 *Ruta graveolens*. (Author: Matias Villacís-Luzuriaga. Cuenca, Ecuador.)

PHYTOCHEMISTRY OF PLANT

Chemical composition varies according to the method applied. The main constituents of *R. graveolens* oil were isolated by preparative GC and further characterized by GC-QTOF-MS and NMR: 2-undecanol, 2-acetate (19.2%), 2-undecanol, 2-methyl butyl ester (8.9%), 1,2,3-trimethylcyclohexane (7.4%) and 2-undecanol, propyl ester (5.0%), 2-octanol, 2-methyl-butyl (5.0%) (Li et al. 2020).

GC-FID and GC-MS analysis applied to the oil extracted from the leaves identified 2-nonanone (39.17%) and 2-undecanone (47.21%) as the principal compounds (França Orlanda and Nascimento 2015).

Chemical composition elucidated by GC and GC-MS determined undecanone-2 (43.66%), 2-nonanone (16.09%), 2-acetoxy tetradecanone (14.49%), and nonyl cyclopropanecarboxylate (9.22%), as the main constituents of the oil extracted from the aerial parts (1.29% yield) (Reddy and Al-Rajab 2016).

BIOLOGICAL ACTIVITIES

R. graveolens directly applied to the skin can cause photodermatitis (Wessner et al. 1999). The essential oil exhibited antibacterial activity against *Bacillus cereus, Enterococcus faecalis, Proteus mirabilis, Micrococcus flavus, Staphylococcus aureus, Escherichia coli, Pseudomona aeruginosa* and *Helicobacter pylori* (França Orlanda and Nascimento 2015, 104–105), (Boughendjioua 2019). The described antimicrobial activity includes fungi (*Candida albicans*) and resistant species such as methicillin-resistant *Staphylococcus aureus* (Reddy and Al-Rajab 2016).

SANTA MARÍA

Botanical name: *Tanacetum parthenium* (L.) Sch. Bip.
Other local name: hierba de Santa María

BOTANICAL INFORMATION AND GEOGRAPHICAL DISTRIBUTION

Family: Asteraceae. It is a perennial, aromatic shrub reaching up to 1 meter in height. It grows in humid environments at altitudes of 3,200 meters (Molina Vélez 2008). Figure 6.14 shows a *T. parthenium* botanical specimen.

TRADITIONAL MEDICINE USES

The plant is used to treat cough, flu, migraine, rheumatic and stomach pain, toothache, fever, menstrual and labor disorders, and is an emmenagogue, to prevent miscarriage, and as an insecticide (Pareek et al. 2011).

PHYTOCHEMISTRY OF PLANT

The major constituents of *T. parthenium* were elucidated through GC and GC-MS: camphor (56.9%), camphene (12.7%), p-cymene (5.2%) and bornyl acetate (4.6%)

FIGURE 6.14 *Tanacetum parthenium*. (Author: Matias Villacís-Luzuriaga. Cuenca, Ecuador.)

(Akpulat et al. 2005). During the flowering stage, bornyl isovalerate, borneol, and β-eudesmol were also identified in the essential oil (Mohsenzadeh et al. 2011).

BIOLOGICAL ACTIVITIES

The essential oil exhibited antibacterial activity against *Staphylococcus subtilis*, *Enterobacter aerogenes*, *Bacillus subtilis* and methicillin-resistant *Staphylococcus aureus* (Mohsenzadeh et al. 2011).

Sauco Negro

Botanical name: *Cestrum peruvianum* Willd. ex Roem. & Schult.
Other local names: sauco, hierba santa

BOTANICAL INFORMATION AND GEOGRAPHICAL DISTRIBUTION

Family: Solanaceae. *Cestrum peruvianum* is a native shrub reaching from 1–4 meters. It grows at 1,000–4,000 meters of altitude (Minga Ochoa and Verdugo Navas 2016). Figure 6.15 displays a *C. peruvianum* botanical specimen.

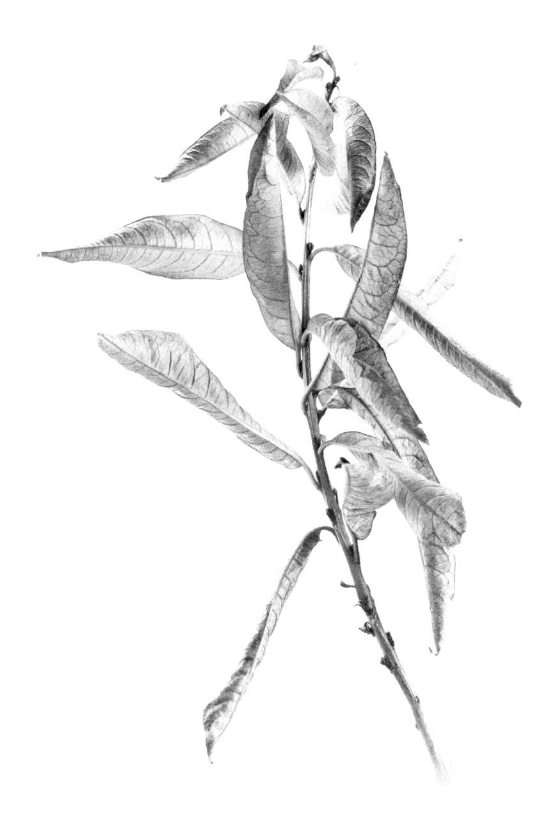

FIGURE 6.15 *Cestrum peruvianum*. (Author: Matias Villacís-Luzuriaga. Cuenca, Ecuador.)

TRADITIONAL MEDICINE USES

The leaves help to control fever by rubbing them on the skin (Ríos et al. 2007). A poultice of the leaves is employed to alleviate muscle and joint pain (Minga Ochoa and Verdugo Navas 2016).

PHYTOCHEMISTRY OF PLANT

To date, no chemical composition of the plant has been described.

BIOLOGICAL ACTIVITIES

No biological activity has been described to date for this plant.

Trinitaria

Botanical name: *Otholobium mexicanum* (L.f.) J. W. Grimes
Other local names: huallua, chanchilva

BOTANICAL INFORMATION AND GEOGRAPHICAL DISTRIBUTION

Family: Fabaceae. *Otholobium mexicanum* is a native shrub that grows up to 2.5 meters, near riverbanks (Molina Vélez 2008). Figure 6.16 presents an *O. mexicanum* botanical specimen.

TRADITIONAL MEDICINE USES

An infusion of this plant is used to treat stomachache and indigestion (Ríos et al. 2007).

PHYTOCHEMISTRY OF PLANT

The phenols: bakuchiol and 3-hydroxybakuchiol, and the isoflavones glycosides: daidzin and genistin, have been identified in *O. mexicanum* (Suárez et al. 2017).

FIGURE 6.16 *Otholobium mexicanum*. (Author: Matias Villacís-Luzuriaga. Cuenca, Ecuador.)

BIOLOGICAL ACTIVITIES

No biological activity has been described to date for this plant.

VERBENA

Botanical name: *Verbena litoralis* Kunth
Other local names: verbena de litoral, verbena de montaña, huarmi verbena, hierba buena, hierba mora de costa

BOTANICAL INFORMATION AND GEOGRAPHIC DISTRIBUTION

Family: Verbenaceae. It is a perennial, native herb from South America. It grows up to 2 meters in height. It develops in tropical to template climates up to 2,000–3,000 meters of altitude (Molina Vélez 2008). Figure 6.17 exhibits a *V. litoralis* botanical specimen.

FIGURE 6.17 *Verbena litoralis*. (Author: Matias Villacís-Luzuriaga. Cuenca, Ecuador.)

TRADITIONAL MEDICINE USES

The juice from the crushed leaves is placed on the forehead to relieve headache. It is also used for wound healing and skin infections. The abdomen is rubbed with twigs and stems for stomach inflammation (Ríos et al. 2007).

PHYTOCHEMISTRY OF PLANT

A diversity of compounds such as: triterpenes, iridoids glycosides, flavonoids, phenylpropanoids-derived, phenylethanoid-derived, cinnamic acid-derived were chemically characterized by UHPLC-ESI-HRMS (de Lima et al. 2018).

BIOLOGICAL ACTIVITIES

V. litoralis showed antinociceptive and antioxidant activity (Braga et al. 2012).

CONCLUSIONS

The "limpia" is a traditional ceremony involving the use of aromatic medicinal plants. The healer uses them as therapeutic tools based on the belief that the intense fragrance of plants attracts evil energy ("mal"). Therefore, the pharmacological activity of this traditional practice seems to be associated with the essential oils of the plants and the topically active compounds. There is evidence of pharmacological activity of essential oils and their main constituents triggered via inhalation or topical, as in the "limpia" ceremony.

Among the described plants, there are comparable pharmacological properties proved by scientific evidence. Such is the case of the anti-inflammatory activity of *Alnus acuminata, Ambrosia arborescens, Baccharis latifolia, Eucalyptus globulus, Brugmansia arborea, Bursera graveolens, Rosmarinus officinalis*; the anti-cancer action of *Ambrosia arborescens, Bursera graveolens, Rosmarinus officinalis*; the antibacterial properties of *Baccharis latifolia, Eucalyptus globulus, Solanum nigrescens, Bursera graveolens, Minthostachys mollis, Rosmarinus officinalis, Ruta graveolens, Tanacetum parthenium*; the antifungal activity of *Baccharis latifolia, Tagetes terniflora, Eucalyptus globulus, Solanum nigrescens, Rosmarinus officinalis, Ruta graveolens*; the wound healing action of *Eucalyptus globulus, Rosmarinus officinalis*; the antioxidant activity of *Eucalyptus globulus, Bursera graveolens, Rosmarinus officinalis, Verbena litoralis*; and the spasmolytic properties of *Eucalyptus globulus, Brugmansia arborea, Bursera graveolens* and *Rosmarinus officinalis*. Thus, the "limpia" ceremony may possess anti-inflammatory, anti-cancer, antibacterial, antifungal, wound healing, antioxidant, and spasmolytic properties based on the active compounds of the plants used in this traditional practice. Toxicity, potential synergism, or antagonism of the active compounds involved in this traditional practice should be further studied.

Andean traditional medicine is a relevant source of phytotherapeutic information. Ideally, these properties should be scientifically confirmed or discarded to encourage the rational use of medicinal plants.

REFERENCES

Abad, M. J., A. L. Bessa, B. Ballarin, O. Aragón, E. Gonzales, and P. Bermejo. 2006. "Anti-Inflammatory Activity of Four Bolivian Baccharis Species (Compositae)." *Journal of Ethnopharmacology* 103 (3). Elsevier: 338–344. doi:10.1016/j.jep.2005.08.024

Aguilar, María I., Ricardo Rovelo, Juan G. Verjan, Oscar Illescas, Ana E. Baeza, Marcela de La Fuente, Ileana Avila, and Andrés Navarrete. 2011. "Anti-Inflammatory Activities, Triterpenoids, and Diarylheptanoids of Alnus Acuminata Ssp. Arguta." *Pharmaceutical Biology* 49 (10). Taylor & Francis: 1052–1057. doi:10.3109/13880209.2011.564634

Akpulat, H. Askin, Bektas Tepe, Atalay Sokmen, Dimitra Daferera, and Moschos Polissiou. 2005. "Composition of the Essential Oils of Tanacetum Argyrophyllum (C. Koch) Tvzel. Var. Argyrophyllum and Tanacetum Parthenium (L.) Schultz Bip. (Asteraceae) from Turkey." *Biochemical Systematics and Ecology* 33 (5). Elsevier Ltd: 511–516. doi:10.1016/j.bse.2004.10.006

Armijos, Chabaco, Iuliana Cota, and Silvia González. 2014. "Traditional Medicine Applied by the Saraguro Yachakkuna: A Preliminary Approach to the Use of Sacred and Psychoactive Plant Species in the Southern Region of Ecuador." *Journal of Ethnobiology and Ethnomedicine* 10 (1). BioMed Central Ltd.: 26. doi:10.1186/1746-4269-10-26

Bachheti, R. K., A. Joshi, and Arjun Singh. 2003. "Oil Content Variation and Antimicrobial Activity of Eucalyptus Leaves Oils of Three Different Species of Dehradun, Uttarakhand, India." *International Journal of ChemTech Research* 3 (2). https://www.researchgate.net/publication/274710903_Oil_Content_variation_and_Antimicrobial_activity_of_Eucalyptus_leaves_oils_of_three_different_Species_of_Dehradun_Uttarakhand_India

Bautista-Valarezo, Estefanía, Víctor Duque, Adriana Elizabeth Verdugo Sánchez, Viviana Dávalos-Batallas, Nele R M Michels, Kristin Hendrickx, and Veronique Verhoeven. 2020. "Towards an Indigenous Definition of Health: An Explorative Study to Understand the indigenous Ecuadorian People's Health and illness Concepts." *International Journal for Equity in Health* 19 (101). doi:10.1186/s12939-020-1142-8

Benny, Anju, and Jaya Thomas. 2019. "Essential Oils as Treatment Strategy for Alzheimer's Disease: Current and Future Perspectives." *Planta Medica* 85 (3). Georg Thieme Verlag: 239–248. doi:10.1055/a-0758-0188

Borges, Raphaelle Sousa, Brenda Lorena Sánchez Ortiz, Arlindo César Matias Pereira, Hady Keita, and José Carlos Tavares Carvalho. 2019. "Rosmarinus Officinalis Essential Oil: A Review of Its Phytochemistry, Anti-Inflammatory Activity, and Mechanisms of Action Involved." *Journal of Ethnopharmacology* 229 (January). Elsevier Ireland Ltd: 29–45. doi:10.1016/j.jep.2018.09.038

Boughendjioua, Hicham. 2019. "Yield, Chemical Composition and Antibacterial Activity of Ruta Chalepensis L. Essential Oil Growing Spontaneously in Algeria." *Pharmacy & Pharmacology International Journal* 7 (1). MedCrave Group, LLC. doi:10.15406/ppij.2019.07.00230

Braga, Virgínia F., Giselle C. Mendes, Raphael T. R. Oliveira, Carla Q. G. Soares, Cristiano F. Resende, Leandro C. Pinto, Reinaldo de Santana, Lyderson F. Viccini, Nádia R. B. Raposo, and Paulo H. P. Peixoto. 2012. "Micropropagation, Antinociceptive and Antioxidant Activities of Extracts of Verbena Litoralis Kunth (Verbenaceae)." *Anais Da Academia Brasileira de Ciencias* 84 (1). An Acad Bras Cienc: 139–147. doi:10.1590/S0001–37652012000100014

Brochot, Amandine, Angèle Guilbot, Laïla Haddioui, and Christine Roques. 2017. "Antibacterial, Antifungal, and Antiviral Effects of Three Essential Oil Blends." *MicrobiologyOpen* 6 (4). Blackwell Publishing Ltd: 459. doi:10.1002/mbo3.459

Bussmann, Rainer W., and Douglas Sharon. 2009. "Shadows of the Colonial Past—Diverging Plant Use in Northern Peru and Southern Ecuador." *Journal of Ethnobiology and Ethnomedicine* 5 (1). BioMed Central: 1–17. doi:10.1186/1746-4269-5-4

Caceres, Armando, Alma v Alvarez, Ana E. Ovando, and Blanca E. Samayoa. 1991. "Plants Used in Guatemala for the Treatment of Respiratory Diseases. 1. Screening of 68 Plants against Gram-Positive Bacteria." *Journal of Ethnopharmacology* 31 (2). Elsevier: 193–208. doi:10.1016/0378–8741(91)90005-X

Cano de Terrones, Teresa. 2014. "Caracterización de Una Espironolactona Sesquiterpénica A-Metilénica Obtenida de Ambrosia Arborescens Miller y Evaluación de Su Actividad Biológica En Tripanozoma Cruzi." *Rev Soc Quím Perú* 80 (2)

Cárdenas, Alexander Ramírez, Gustavo Isaza Mejía, Jorge E Pérez Cárdenas, and Maby M Martínez Garzón. 2017. "Estudio Fitoquímico Preliminar y Evaluación de La Actividad Antibacteriana Del Solanum Dolichosepalum Bitter (Frutillo) Preliminary Phytochemical Study and Evaluation of the Antibacterial Activity of Solanum Dolichosepalum Bitter (Frutillo)." *Revista Cubana de Plantas Medicinales* 22 (1). http://scielo.sld.cu

Carvalho, M. Fernanda N.N., Ana M. Botelho do Rego, Adelino M. Galvão, Rudolf Herrmann, and Fernanda Marques. 2018. "Search for Cytotoxic Compounds against Ovarian Cancer Cells: Synthesis, Characterization and Assessment of the Activity of New Camphor Carboxylate and Camphor Carboxamide Silver Complexes." *Journal of Inorganic Biochemistry* 188 (November). Elsevier Inc.: 88–95. doi:10.1016/j.jinorgbio.2018.08.011

Cavender, Anthony P., and Manuel Albán. 2009. "The Use of Magical Plants by Curanderos in the Ecuador Highlands." *Journal of Ethnobiology and Ethnomedicine* 5 (3). BioMed Central: 1–9. doi:10.1186/1746-4269-5-3

Cruz-Carrillo, Anastasia, Carlos Rodríguez Molano, and Carlos Ortíz-Lopez. 2011. "Efecto Insecticida in Vitro Del Extracto Etanólico de Algunas Plantas Sobre La Mosca Adulta Haematobia Irritans In Vitro Insecticidal Effect of the Ethanolic Extract from Some Plants on the Adult Fly Haematobia Irritans." *Revista Cubana de Plantas Medicinales* 16 (3): 216–226. http://scielo.sld.cuhttp://scielo.sld.cu217

de Leo, Marinella, Mariela Beatriz Vera Saltos, Blanca Fabiola Naranjo Puente, Nunziatina de Tommasi, and Alessandra Braca. 2010. "Sesquiterpenes and Diterpenes from Ambrosia Arborescens." *Phytochemistry* 71 (7): 804–809. doi:10.1016/j.phytochem.2010.02.002

de Lima, R., C.G. Guex, A.R.H. da Silva, C.L. Lhamas, K.L. dos Santos Moreira, R. Casoti, R.C. Dornelles, et al. 2018. "Acute and Subacute Toxicity and Chemical Constituents of the Hydroethanolic Extract of Verbena Litoralis Kunth." *Journal of Ethnopharmacology* 224 (October). J Ethnopharmacol: 76–84. doi:10.1016/J.JEP.2018.05.012

de Oliveira, Jonatas Rafael, Samira Esteves Afonso Camargo, and Luciane Dias de Oliveira. 2019. "Rosmarinus Officinalis L. (Rosemary) as Therapeutic and Prophylactic Agent." *Journal of Biomedical Science*. BioMed Central Ltd. doi:10.1186/s12929-019-0499-8

Dey, Baishakhi, and Analava Mitra. 2013. "Chemo-Profiling of Eucalyptus and Study of Its Hypoglycemic Potential." *World Journal of Diabetes* 4 (5). Baishideng Publishing Group Inc.: 170. doi:10.4239/wjd.v4.i5.170

Dhakad, Ashok K., Vijay v. Pandey, Sobia Beg, Janhvi M. Rawat, and Avtar Singh. 2018. "Biological, Medicinal and Toxicological Significance of Eucalyptus Leaf Essential Oil: A Review." *Journal of the Science of Food and Agriculture* 98 (3). John Wiley and Sons Ltd.: 833–848. doi:10.1002/jsfa.8600

Feriotto, Giordana, Nicola Marchetti, Valentina Costa, Simone Beninati, Federico Tagliati, and Carlo Mischiati. 2018. "Chemical Composition of Essential Oils from Thymus Vulgaris, Cymbopogon Citratus, and Rosmarinus Officinalis, and Their Effects on the HIV-1 Tat Protein Function." *Chemistry and Biodiversity* 15 (2). Wiley-VCH Verlag: e1700436. doi:10.1002/cbdv.201700436

Fernández-Ruiz, Mashiel, Liris Yepes-Fuentes, Irina Tirado-Ballestas, and Mauricio Orozco. 2018. "Actividad Repelente Del Aceite Esencial de Bursera Graveolens Jacq. Ex L., Frente Tribolium Castaneum Herbst, 1797 (Coleoptera: Tenebrionidae)." *Anales de Biología* 40: 87–93. doi:10.6018/analesbio.40.10

Fon-Fay, Flor M., Jorge A. Pino, Ivones Hernández, Idania Rodeiro, and Miguel D. Fernández. 2019. "Chemical Composition and Antioxidant Activity of Bursera Graveolens (Kunth) Triana et Planch Essential Oil from Manabí, Ecuador." *Journal of Essential Oil Research* 31 (3). Taylor and Francis Inc.: 211–216. doi:10.1080/10412905.2018.1564381

França Orlanda, J. F., and A. R. Nascimento. 2015. "Chemical Composition and Antibacterial Activity of Ruta Graveolens L. (Rutaceae) Volatile Oils, from São Luís, Maranhão, Brazil." *South African Journal of Botany* 99 (July). Elsevier: 103–106. doi:10.1016/j.sajb.2015.03.198

Galvez, Carolina E., Cristina M. Jimenez, Analía de los A. Gomez, Emilio F. Lizarraga, and Diego A. Sampietro. 2018. "Chemical Composition and Antifungal Activity of Essential Oils from Senecio Nutans, Senecio Viridis, Tagetes Terniflora and Aloysia Gratissima against Toxigenic Aspergillus and Fusarium Species." *Natural Product Research* 34 (10). Taylor & Francis: 1442–1445. doi:10.1080/14786419.2018.1511555

Girón, Lidia M., Gustavo A. Aguilar, Armando Cáceres, and Gerardo L. Arroyo. 1988. "Anticandidal Activity of Plants Used for the Treatment of Vaginitis in Guatemala and Clinical Trial of a Solanum Nigrescens Preparation." *Journal of Ethnopharmacology* 22 (3). Elsevier: 307–313. doi:10.1016/0378–8741(88)90241–3

Houghton, Peter J. 1995. "The Role of Plants in Traditional Medicine and Current Therapy." *Journal of Alternative and Complementary Medicine* 1 (2). J Altern Complement Med: 131–143. doi:10.1089/acm.1995.1.131

Ikawati, Zullies, Subagus Wahyuono, and Kazutaka Maeyama. 2001. "Screening of Several Indonesian Medicinal Plants for Their Inhibitory Effect on Histamine Release from RBL-2H3 Cells." *Journal of Ethnopharmacology* 75 (2–3). J Ethnopharmacol: 249–256. doi:10.1016/S0378–8741(01)00201-X

Jiang, Yang, Nan Wu, Yu Jie Fu, Wei Wang, Meng Luo, Chun Jian Zhao, Yuan Gang Zu, and Xiao Lei Liu. 2011. "Chemical Composition and Antimicrobial Activity of the Essential Oil of Rosemary." *Environmental Toxicology and Pharmacology* 32 (1). Environ Toxicol Pharmacol: 63–68. doi:10.1016/j.etap.2011.03.011

Juergens, U. R., M. Stöber, and H. Vetter. 1998. "Inhibition of Cytokine Production and Arachidonic Acid Metabolism by Eucalyptol (1.8-Cineole) in Human Blood Monocytes in Vitro." *European Journal of Medical Research* 3 (11): 508–510. https://europepmc.org/article/med/9810029

Jun, Yang Suk, Purum Kang, Sun Seek Min, Jeong Min Lee, Hyo Keun Kim, and Geun Hee Seol. 2013. "Effect of Eucalyptus Oil Inhalation on Pain and Inflammatory Responses after Total Knee Replacement: A Randomized Clinical Trial." *Evidence-Based Complementary and Alternative Medicine* 2013. Hindawi Limited: 1–7. doi:10.1155/2013/502727

Kim, Hyoung-Geun, Davin Jang, Young Sung Jung, Hyun-Ji Oh, Seon Min Oh, Yeong-Geun Lee, Se Chan Kang, Dae-Ok Kim, and Dae Young Lee and Nam-In Baek. 2020. "Anti-Inflammatory Effect of Flavonoids from Brugmansia Arborea L. Flowers." *J. Microbiol. Biotechnol.* 30 (2). Korean Society for Microbiology and Biotechnology: 163–171. doi:10.4014/JMB.1907.07058

Labib, Rola M., Iriny M. Ayoub, Haidy E. Michel, Mina Mehanny, Verena Kamil, Meryl Hany, Mirette Magdy, Aya Moataz, Boula Maged, and Ahmed Mohamed. 2019. "Appraisal on the Wound Healing Potential of Melaleuca Alternifolia and Rosmarinus Officinalis L. Essential Oil-Loaded Chitosan Topical Preparations." *PLoS ONE* 14 (9). Public Library of Science. doi:10.1371/journal.pone.0219561

Li, Ying, Guangyao Dong, Xi Bai, Aoken Aimila, Xiaohui Bai, Maitinuer Maiwulanjiang, and H. A. Aisa. 2020. "Separation and Qualitative Study of Ruta Graveolens L. Essential Oil Components by Prep-GC, GC-QTOF-MS and NMR." *Natural Product Research* 35 (21). Taylor and Francis Ltd.: 4202–4205. doi:10.1080/14786419.2020.1756798

Lin, Yu Hsing, Jong Tar Kuo, Yun Yu Chen, K. J. Senthil Kumar, Chiu Ping Lo, Chin Chung Lin, and Sheng Yang Wang. 2018. "Immunomodulatory Effects of the Stout Camphor Medicinal Mushroom, Taiwanofungus Camphoratus (Agaricomycetes)—Based Health Food Product in Mice." *International Journal of Medicinal Mushrooms* 20 (9). Begell House Inc.: 849–858. doi:10.1615/IntJMedMushrooms.2018027389

Loayza, Ingrid, David Abujder, Rosemary Aranda, Jasmin Jakupovic, Guy Collin, Hélène Deslauriers, and France Ida Jean. 1995. "Essential Oils of Baccharis Salicifolia, B. Latifolia and B. Dracunculifolia." *Phytochemistry* 38 (2). Pergamon: 381–389. doi:10.1016/0031–9422(94)00628–7

Manoharan, Ranjith Kumar, Jin Hyung Lee, and Jintae Lee. 2017. "Antibiofilm and Antihyphal Activities of Cedar Leaf Essential Oil, Camphor, and Fenchone Derivatives against Candida Albicans." *Frontiers in Microbiology* 8 (Aug). Frontiers Media S.A.: 1476. doi:10.3389/fmicb.2017.01476

Manzano Santana, Patricia, Migdalia Miranda, Yamilet Gutiérrez, Gastón García, Tulio Orellana, and Andrea Orellana. 2009. "Efecto Antiinflamatorio y Composición Química Del Aceite de Ramas de Bursera Graveolens Triana & Planch. (Palo Santo) de Ecuador Antinflammatory Effect and Chemical Composition of Bursera Graveolens Triana & Planch. Branch Oil (Palo Santo) from Ecuador." *Revista Cubana de Plantas Medicinales* 14 (3): 45–53.

Mathez-Stiefel, Sahra, Sébastien Boillat, and Stephan Rist. 2007. "Promoting the Diversity of Worldviews: An Ontological Approach to Biocultural Diversity." In *Endogenous Development and Bio-Cultural Diversity. The Interplay of Worldviews, Globalization and Locality*, 67–81. Leusden: ETC/COMPAS

Minga, Danilo. 2000. *Árboles y Arbustos Del Bosque de Mazán*. Cuenca: ETAPA. doi:10.13140/2.1.3829.8403

Minga Ochoa, Danilo, and Adolfo Verdugo Navas. 2016. *Árboles y Arbustos de Los Ríos de Cuenca, Azuay, Ecuador*. Cuenca: Don Bosco. https://books.google.com/books/about/%C3%81rboles_y_arbustos_de_los_r%C3%ADos_de_Cuen.html?hl=es&id=aRiAAQAACAAJ

MIlhau, Gullhem, Alexis Valentin, Françise Benoit, Michèle Mallié, Jean Marie Bastide, Yves Pélissier, and Jean Marie Bessière. 1997. "In Vitro Antimalarial Activity of Eight Essential Oils." *Journal of Essential Oil Research* 9 (3). Taylor & Francis Group: 329–333. doi:10.1080/10412905.1997.10554252

Mohamed, Gamal A., and Sabrin R. M. Ibrahim. 2007. "Eucalyptone G, a New Phloroglucinol Derivative and Other Constituents from Eucalyptus Globulus Labill." *Arkivoc* 2007 (15). Arkat: 281–291. doi:10.3998/ark.5550190.0008.f27

Mohsenzadeh, F., A. Chehregani, and H. Amiri. 2011. "Chemical Composition, Antibacterial Activity and Cytotoxicity of Essential Oils of Tanacetum Parthenium in Different Developmental Stages." *Pharmaceutical Biology* 49 (9). Pharm Biol: 920–926. doi:10.3109/13880209.2011.556650

Molina Vélez, Magdalena. 2008. *Fitoterapia*. Cuenca: Casa de la Cultura Ecuatoriana "Benjamín Carrión" Núcleo del Azuay. https://www.goodreads.com/book/show/53418104-fitoterapia

Monzote, Lianet, Gabrielle M Hill, Armando Cuellar, Ramón Scull, and William N Setzer. 2021. "Chemical Composition and Anti-Proliferative Properties of Bursera Graveolens Essential Oil." *Natural Product Communications* 7 (11): 1531–1534. https://pubmed.ncbi.nlm.nih.gov/23285824/

Mora, Flor D, María Araque, Luis B Rojas, Rosslyn Ramirez, Bladimiro Silva, and Alfredo Usubillaga. 2009. "Chemical Composition and in Vitro Antibacterial Activity of the Essential Oil of Minthostachys Mollis (Kunth) Griseb Vaught from the Venezuelan Andes." *Nat Prod Commun* 4 (7): 997–1000. https://pubmed.ncbi.nlm.nih.gov/19731611/

Pareek, Anil, Manish Suthar, Garvendra S. Rathore, and Vijay Bansal. 2011. "Feverfew (Tanacetum Parthenium L.): A Systematic Review." *Pharmacognosy Reviews*. Pharmacogn Rev. doi:10.4103/0973–7847.79105

Pellegrini, María C., Rosa M. Alonso-Salces, María L. Umpierrez, Carmen Rossini, and Sandra R. Fuselli. 2017. "Chemical Composition, Antimicrobial Activity, and Mode of Action of Essential Oils against Paenibacillus Larvae, Etiological Agent of American Foulbrood on Apis Mellifera." *Chemistry and Biodiversity* 14 (4). Wiley-VCH Verlag: e1600382. doi:10.1002/cbdv.201600382

Peralta Gómez, Susana, Aarland Rayn Clarenc, Maria G. Campos Lara, María Elena Jiménez Lara, and José Alberto, Mendoza Espinoza. 2013. "Toxicity Analysis, Phytochemical and Pharmacological Study of the Plant Known as Mora Herb, Collected at the Environmental Education Center of Yautlica (CEA-Yautlica)." *Asian Journal of Plant Sciences* 12 (4): 159–164. doi:10.3923/AJPS.2013.159.164

Pino, José, and Rafael Alvis. 2009. "Effects of Brugmansia Arborea (L.) Lagerheim (Solanacea) in the Male Reproductive System of Mouse." *Rev. Peru. Biol.* 15 (2): 125–127. http://sisbib.unmsm.edu.pe/BVRevistas/biologia/biologiaNEW.htmRev.peru.biol.15

Quezada, Alberto, César Hermida, Gustavo Vega, Ernesto Cañizares, and Jorge Hermida. 1992. *La Práctica Médica Traditional [Traditional Medical Practice]*. 2nd ed. Cuenca: Universidad de Cuenca.

Rahbardar, Mahboobeh Ghasemzadeh, and Hossein Hosseinzadeh. 2020. "Therapeutic Effects of Rosemary (Rosmarinus Officinalis L.) and Its Active Constituents on Nervous System Disorders." *Iranian Journal of Basic Medical Sciences* 23 (9). Mashhad University of Medical Sciences: 1100–1112. doi:10.22038/ijbms.2020.45269.10541

Rašković, Aleksandar, Isidora Milanović, Nebojša Pavlović, Tatjana Ćebović, Saša Vukmirović, and Momir Mikov. 2014. "Antioxidant Activity of Rosemary (Rosmarinus Officinalis L.) Essential Oil and Its Hepatoprotective Potential." *BMC Complementary and Alternative Medicine* 14 (July). BioMed Central Ltd. doi:10.1186/1472-6882-14-225

Reddy, Desam Nagarjuna, and Abdul Jabbar Al-Rajab. 2016. " Chemical Composition, Antibacterial and Antifungal Activities of Ruta Graveolens L . Volatile Oils ." *Cogent Chemistry* 2 (1). Informa UK Limited: 1220055. doi:10.1080/23312009.2016.1220055

Ríos, Montserrat, M.J. Koziol, H. Borgtoft Pedersen, and G. Granda, eds. 2007. *Plantas Útiles Del Ecuador: Aplicaciones, Retos y Perspectivas*. Quito: Abya-Yala.

Russo, Ricardo O. 1990. "Evaluating Alnus Acuminata as a Component in Agroforestry Systems." *Agroforestry Systems* 10 (3). Kluwer Academic Publishers: 241–252. doi:10.1007/BF00122914

Sadlon, Angela E., and Davis W. Lamson. 2010. "Immune-Modifying and Antimicrobial Effects of Eucalyptus Oil and Simple Inhalation Devices." *Alternative Medicine Review* 15 (1). Thorne Research Inc.: 33–43. https://go.gale.com/ps/i.do?p=AONE&sw=w&issn=10895159&v=2.1&it=r&id=GALE%7CA225739691&sid=googleScholar&linkaccess=fulltext

Said, Z. S., Sakina Slimani, Hocine Remini, Hayat Idir-Himed, J. Mazauric, Khodir Madani, and L. Boulekbache-Makhlouf. 2016. "Phytochemical Analysis and Antioxidant Activity of Eucalyptus Globulus : A Comparative Study between Fruits and Leaves Extracts." *Journal of Chemical Engineering & Bioanalytical Chemistry* 1 (1). https://www.researchgate.net/publication/291074463

Salama, Ahmed M, and Inés Yamile Avendaño. 2021. "Actividad Antiinflamatoria de D-Amirona y 4', 7-Dimetoxiapigenina Aislados de Alnus Acuminata." Accessed June 15. www.farmacia.unal.edu.co

Sampath, Sowndarya, Sangeetha Subramani, Sridevi Janardhanam, Preethi Subramani, Arun Yuvaraj, and Rose Chellan. 2018. "Bioactive Compound 1,8-Cineole Selectively Induces G2/M Arrest in A431 Cells through the Upregulation of the P53 Signaling Pathway and Molecular Docking Studies." *Phytomedicine* 46 (July). Elsevier GmbH: 57–68. doi:10.1016/j.phymed.2018.04.007

Sánchez-Chopa, Carolina, and Lilian R Descamps. 2012. "Composition and Biological Activity of Essential Oils against Metopolophium Dirhodum (Hemiptera: Aphididae) Cereal Crop Pest." *Pest Management Science* 68 (11). Pest Manag Sci: 1492–1500. doi:10.1002/PS.3334

Sánchez-Recillas, A., S. L. Aragón-Castillo, A. L. Arroyo-Herrera, J.A. Araujo-León, and R. R. Ortiz-Andrade. 2020. "Efecto Espasmolitico y Antibacteriano de La Especie Bursera Graveolens (Kunth)." *Polibotánica* 0 (49). Escuela Nacional de Ciencias Biologicas: 135–147. doi:10.18387/polibotanica.49.9

Schmidt-Lebuhn, AN. 2008. "Ethnobotany, Biochemistry and Pharmacology of Minthostachys (Lamiaceae)." *Journal of Ethnopharmacology* 118 (3). J Ethnopharmacol: 343–353. doi:10.1016/J.JEP.2008.05.030

Sequeda-Castañeda, Luis, Crispin Célis, and Pilar Luengas-Caicedo. 2015. "Phytochemical and Therapeutic Use of Baccharis Latifolia (Ruiz & Pav.) Pers. (Asteraceae)." *Archives* 2: 14–17.

Sharma, Arun Dev, and Inderjeet Kaur. 2020. "Eucalyptol (1,8 Cineole) from Eucalyptus Essential Oil a Potential Inhibitor of COVID 19 Corona Virus Infection by Molecular Docking Studies." *Preprints*, March. Preprints. doi:10.20944/preprints202003.0455.v1

Sotillo, Wendy Soria, Rodrigo Villagomez, Sandra Smiljanic, Xiaoli Huang, Atena Malakpour, Sebastian Kempengren, Gloria Rodrigo, Giovanna Almanza, Olov Sterner, and Stina Oredsson. 2017. "Anti-Cancer Stem Cell Activity of a Sesquiterpene Lactone Isolated from Ambrosia Arborescens and of a Synthetic Derivative." *PLoS ONE* 12 (9). Public Library of Science. doi:10.1371/journal.pone.0184304

Suárez, Alírica I., Zaw Min Thu, Jorge Ramírez, Diana León, Luis Cartuche, Chabaco Armijos, and Geovanni Vidari. 2017. "Main Constituents and Antidiabetic Properties of Otholobium Mexicanum:" *Natural Products Communications* 12 (4). SAGE PublicationsSage CA: Los Angeles, CA: 533–535. doi:10.1177/1934578X1701200418

Svensson, Daniel, Maribel Lozano, Giovanna R. Almanza, Bengt Olof Nilsson, Olov Sterner, and Rodrigo Villagomez. 2018. "Sesquiterpene Lactones from Ambrosia Arborescens Mill. Inhibit pro-Inflammatory Cytokine Expression and Modulate NF-KB Signaling in Human Skin Cells." *Phytomedicine* 50 (November). Elsevier GmbH: 118–126. doi:10.1016/j.phymed.2018.04.011

Taur, D. J., V. B. Kulkarni, and R. Y. Patil. 2010. "Chromatographic Evaluation and Anthelmintic Activity of Eucalyptus Globulus Oil." *Pharmacognosy Research* 2 (3). Wolters Kluwer—Medknow Publications: 125–127. doi:10.4103/0974-8490.65504

Tene, Vicente, Omar Malagón, Paola Vita Finzi, Giovanni Vidari, Chabaco Armijos, and Tomás Zaragoza. 2007. "An Ethnobotanical Survey of Medicinal Plants Used in Loja and Zamora-Chinchipe, Ecuador." *Journal of Ethnopharmacology* 111 (1). Elsevier: 63–81. doi:10.1016/j.jep.2006.10.032

Tomen, Ibrahim, Fatma Tugce Guragac, Hikmet Keles, Markku Reunanen, and Esra Kupeli-Akkol. 2017. "Characterization and Wound Repair Potential of Essential Oil Eucalyptus Globulus Labill." *Fresenius Environmental Bulletin* 26 (11). Parlar Scientific Publications: 6390–6399.

Valarezo, Eduardo, Marco Rosillo, Luis Cartuche, Omar Malagón, Miguel Meneses, and Vladimir Morocho. 2013. "Chemical Composition, Antifungal and Antibacterial Activity of the Essential Oil from Baccharis Latifolia (Ruiz & Pav.) Pers. (Asteraceae) from Loja, Ecuador." *Journal of Essential Oil Research* 25 (3). Routledge: 233–238. doi: 10.1080/10412905.2013.775679

Villagomez, Rodrigo, Gloria C. Rodrigo, Isidro G. Collado, Marco A. Calzado, Eduardo Muñoz, Björn Akesson, Olov Sterner, Giovanna R. Almanza, and Rui Dong Duan. 2013. "Multiple Anticancer Effects of Damsin and Coronopilin Isolated from Ambrosia Arborescens on Cell Cultures." *Anticancer Research* 33 (9). Anticancer Res: 3799–3806. https://pubmed.ncbi.nlm.nih.gov/24023312/

Villareal, Myra O., Ayumi Ikeya, Kazunori Sasaki, Abdelkarim ben Arfa, Mohamed Neffati, and Hiroko Isoda. 2017. "Anti-Stress and Neuronal Cell Differentiation Induction Effects of Rosmarinus Officinalis L. Essential Oil." *BMC Complementary and Alternative Medicine* 17 (1). BioMed Central Ltd. doi:10.1186/s12906-017-2060-1

Wessner, D., H. Hofmann, and J. Ring. 1999. "Phytophotodermatitis Due to Ruta Graveolens Applied as Protection against Evil Spells." *Contact Dermatitis* 41 (4). Blackwell Munksgaard: 232. doi:10.1111/j.1600–0536.1999.tb06145.x

Young, Don Gary. 2007. "Essential Oil of Bursera Graveolens (Kunth) Triana et Planch from Ecuador." *Article in Journal of Essential Oil Research*. doi:10.1080/10412905.2007.9699322

Zapata, Bibiana, Camilo Durán, Elena Stashenko, Liliana Betancur-Galvis, and Ana Cecilia Mesa-Arango. 2010. "Actividad Antimicótica y Citotóxica de Aceites Esenciales de Plantas de La Familia." *Revista Iberoamericana de Micologia* 27 (2). Rev Iberoam Micol: 101–103. doi:10.1016/j.riam.2010.01.005

Zapata-Maldonado, Christian, Miguel Serrato-Cruz, Emmanuel Ibarra, and Blanca Naranjo-Puente. 2015. "Chemical Compounds of Essential Oil of Tagetes Species of Ecuador." *ECOFARN Journal* 1 (1): 19–26. www.ecorfan.org/republicofnicaragua

Part III

Important Plant Species of Ecuador

This section begins with a general compendium of important medicinal plants for cancer treatment that including native and foreign plants species. Later chapters introduce some of the newest findings on Ecuadorian native plants with importance for conservation, medicine, pharmacological and nutraceutical purposes

DOI: 10.1201/9781003173991-9

7 Medicinal Plants and Epigenetic Mechanisms in Cancer

Andrea Orellana-Manzano, Carlos Ordoñez, Patricia Manzano and Glenda Pilozo

CONTENTS

MEDICINAL PLANTS IN MAMA CANCER

Capsicum chinense

This species, commonly called habanero chili, is of great gastronomic and cultural importance. Traditional medicine is used for various ailments such as headache, rheumatism, night blindness, digestive diseases, gastritis, spondylitis, arthritis and ankylosing spondylitis. These healing properties are related to its content of bioactive components such as capsaicinoids, carotenoids, flavonoids, phenols, alkaloids and vitamins. In addition, its consumption contributes to the prevention of chronic diseases such as cancer and cardiovascular diseases. It also exhibits antimicrobial, anti-inflammatory, antioxidant, antifungal and antiproliferative properties. (Velázquez-Hernández et al. 2021).

DOI: 10.1201/9781003173991-10

The thionine's epigenetics mechanism of the γ-thionine could form transitory pores that allowed the molecule's internalization and led to the apoptosis pathway. The AMP γ-thionine inhibits the Histone deacetylase in the early stage. Besides, regulate negativity the transcription of HDAC 1, 2, 3 in long periods. This could lead to an increase in the global H3K9ac. This state could generate an increase of the oxygen reactive species and activate the apoptosis pathway. Secondly, the defensive thionine decreases the transcript of KDM4A, favoring the increase of the H3K9m2, related with decreased ERα and transcriptional repression of targets genes. AMP γ-thionine inhibits the expression of the codification's gene ERalfa and cyclin D1, suggesting that it could be related to a decreased cellular proliferation. According to Arceo, this could lead to a cytotoxic activity in cancer cells line with the estrogenic response, which could modulate the transcription for the more potent receptor (Martínez 2018).

Vitis vinifera

This is a plant species native to Mediterranean Europe and Central Asia, is highly commercialized, as it is used in the production of juices, wine and other foods. Its main bioactive compound is resveratrol, a stilbenoid polyphenol of pharmacological importance. *In vivo* and *in vitro* tests attribute antioxidant, anti-inflammatory, neuroprotective, anti-aging, anti-obesity and anti-diabetes mellitus properties to prevent cardiovascular diseases and cancer development (Mahanna et al. 2019; Figueira and González 2018; Griñán-Ferré et al. 2021; Zheng and Chen 2021).

The anticancer property of resveratrol is evidenced by its ability to inhibit breast adenocarcinoma through the regulation of microRNA, which induces the process of apoptosis in cells, which could be related to the down-regulation of miRNA 21 and 15, and the up-regulation of miR 125 and 126, being able to modify the genes Pten-AKT-mTOR, FOX 3, IGFB associated with cell migration (Ahmed et al. 2020).

Nigella sativa

This is a species native to India, commonly called Black seed, recognized for its use in gastronomic and ancestral medicine, and is used to combat the adverse effects of asthma, inflammation and arthritis. It is also known for its potential activity against cancer, diabetes, hypertension, epilepsy, ulcers, fatty liver and oxidative stress, attributed to its content of bioactive molecules such as thymol, α-phellandrene, oleic acid, proteins, carbohydrates and – one of the most representative – thymoquinone, which has essential pharmacological activities (Mostafa et al. 2021; Majeed et al. 2021).

In addition, one of its most prominent effects is its ability to inhibit cancer cell growth by intervening in the carcinogenic process. This is possible by modulating the epigenetic features that induce apoptosis of these cancer cells by inhibiting histone deacetylation and DNA methylation or demethylation. Thymoquinone likely causes a reduction in the expression of epigenetic proteins, such as regulating HDACs, which prevents their overexpression, thus inducing histone acetylation and preventing histone deacetylation. Other proteins regulated by thymoquinone include DNMT1, 3A, 3B, G9A, HDAC1,4,9, KDM1B, KMT2A, B, C, D and E (Khan et al. 2019).

Bursera graveolens

B. graveolens, commonly referred to as Palo santo, according to the Dictionary of South American Trees, may also be referred to as *Amyris caranifera, Bursera penicillata, Bursera pubescens, Bursera tatamaco, Elaphrium graveolens, Elaphrium penicillatum, Elaphrium pubescens, Elaphrium tatamaco, Spondias edmonstonei, Terebinthus graveolens and Terebinthus pubescens;* belongs to the Bruceraceae family, native to the coast of Mexico, Colombia, Ecuador, Peru and Brazil (Grandtner and Chevrette 2013).

MEDICINAL PLANTS IN COLON CANCER

Artemisis sieversiana

This is a species native to Asia, Europe, and North America, and is traditionally used to treat fever and infections. The main bioactive compounds that make up the genus Artemisia are terpenoids, flavonoids, coumarins, caffeoylquinic acids, sterols, acetylenes and several compounds with high pharmacological potential, such as limonene, which has recognized anticancer and inhibitory actions on liver tumors. In addition, it has been reported that this species has anthelmintic and insecticidal properties. The essential oil acts as an anti-inflammatory (Adewumi et al. 2020; Pellicer et al. 2018).

It presents cytotoxic activity in cancer cells. This effect may correspond to epigenetic modification in HT-29, HCT15 and COLO 205 lines, in which it has been reported that the ethanol extract of the aerial parts induces apoptosis, DNA damage, and loss of mitochondrial membrane potential and may exert its cytotoxicity through activation of autophagic pathways such as EIF2, phagosome maturation eIF4/p70S6K (Benarba and Pandiella 2018).

Ceratonia siliqua

This species is commonly called carob, is native to the Mediterranean, has a high nutritional and medicinal value, and is traditionally used as a sweetener, diuretic, purgative, laxative to treat diarrhea, as well as for diabetes, herpes, obesity, menorrhagia, killing intestinal parasites, and treating bacterial infections, cough and fevers. Its main constituents are tannins, pectins, hemicellulose, cellulose, nitrogen, mineral elements, sugars, minerals, vitamins and phenolic compounds. These are responsible for nephroprotective, antiatherogenic, antibacterial, antimicrobial, antimicrobial, antimicrobial, antimicrobial, antimicrobial, antimicrobial, antifungal, antidiabetic, antiviral, anti-inflammatory, and antioxidant antifibrotic, antidepressant, antiatherosclerotic, anxiolytic-sedative, antiproliferative, anticancer and antiproliferative properties (Azab 2017).

It has been reported that its polyphenolic extract induces apoptosis in colon adenocarcinoma cells HCT-116 and CT-26. This could be due to the modification of the mitochondrial pathway, which could induce the activation of the enzymes caspase 9, 3, and 7, PARP excision, the increase of p38 MAPK, and phosphorylation of p53, which would explain the initiation of the autophagy process in these cancer cells (Benarba and Pandiella 2018).

MEDICINAL PLANTS IN PROSTATE CANCER

Radix angelicae sinensis

This is a species native to China, known as Danggui, and is traditionally consumed as a dietary supplement or in natural medicine. It is used to treat various ailments such as anemia, hepatitis, ulcerative colitis, dermatosis, neuropathy, cancer, diabetes and nephrosis. It is also used to regulate menstruation, promote blood circulation, relieve pain and constipation. It is also used to treat the heart, spleen, liver and kidneys (Su et al. 2013; Li et al. 2009).

Among its most important bioactive components is Z-Ligustilide. One of its most promising pharmacological effects is its ability to fight prostate cancer. This occurs by modifying epigenetic mechanisms in TRAMP C1 cells that may be related to the restoration of Nrf2 gene expression by decreasing the methylation of its promoter. This could induce mRNA expression and proteins of endogenous target genes such as HO-1, NQO1, and UGT1A1. In addition, it has been reported that treatment with this species can reduce CpG methylation of the Nrf2 gene promoter and decrease AND methylation in the Nrf2 gene region (Su et al. 2013).

Withania somnifera

This is a species native to Asia and South Africa, commonly called Indian ginseng, and is used in traditional medicine as a rejuvenating nutrient that raises the rate of hemoglobin and melanin pigmentation of hair. It is also used for treating rheumatic inflammation, ulcer healing, and as a supplement for pregnant women or older adults. It is anticancer, antituberculosis, anthelmintic, antipyretic and antiulcer. Among its constituents of great pharmacological value are withanolides, sitoindosides and other alkaloids. Several studies have reported immunomodulatory, Th1-specific anti-inflammatory, anticarcinogenic and anticancer activities (Rayees and Malik 2017).

This species can prevent the formation of cancer cells and their proliferation in the development of prostate cancer by gene silencing. This could be because its bioactive component triethylene glycol activates the tumor suppressor proteins p53 and pRB. In addition, it is presumed that the ability to disrupt the binding of the BIR5 protein in the mitotic process could be related to its antitumor activity (Rai et al. 2016).

MEDICINAL PLANTS IN OTHER TYPES OF CANCER

Curcuma longa

This is a medicinal species native to India, traditionally used as a tonic, antidote, anti-inflammatory, antiseptic, preservative, anthelmintic, stimulant, blood purifier, circulatory stimulant, wound healing, liver cleanser, immune system strengthening, thrombosis, indigestion, diabetes, high cholesterol, obesity, arthritis and kidney infection. It is also a condiment and is used to combat liver disorders, skin, cancer, hemorrhoids, asthma, inflammation and leprosy. *Curcuma* longa is constituted by phenolic compounds such as diarylheptanoids, diyrenphenate and isoflavone. Additionally, it is constituted by volatile compounds such as monoterpenes, sesquiterpenes, diterpenes and triterpenoids (Ayati et al. 2019).

TABLE 7.1
Bibliographic Review of the Medicinal Plants and Epigenetics Modification in Different Types of Cancer

Medicinal plant	Origen	Type of Study	Bioactive compound	Epigenetics modifications	Epigenetics regulators	Target	Mechanism	Biological function	Cancer type	References
Curcuma longa	India	Bibliographic review	Polyphenol bisdemethoxyc-urcumin	hypermethylation and DNMT inhibition	NF-kB, Akt, Wnt/β-catenina, AR	Restoring the expression of WIF1	Wnt/β-catenin signaling pathway	Therapeutic or chemopreventive agent	Lung Cancer	Echeverry et al. (2016)
Capsicum chinese	México	Experimental study	AMP γ-tionina	Acetylation	Histone H3K9	Cyclin D1 and HDAC 1/2/3	Increase in acetylation levels in H3K9ac and H3K9me2	Cytotoxic activity in response to estrogens	Breast cancer	Arceo Martínez (2018)
Nigella sativa; family Ranunculaceae	India	Bibliographic review	Thymoquinone	miRNA upregulation: miR34a Downregulation: miR206b-3p, miR146a	mR34a, miR206b-3p and miR146a	RAC1. Necrosis and Oxidative stress. NF-kb	*miR32a upregulation inhibited breast cancer cell metastasis. Downregulation of RAC1 expression and induce actin polymerization and disruption. *miR206b-3p is linked with increased necrosis and oxidative stress. * miRNA146a inhibits inflammatory cytokines through the NF-kB pathway	Anticancer activity	Breast cancer	Khan et al. (2019)

(Continued)

TABLE 7.1 (Continued)

Medicinal plant	Origen	Type of Study	Bioactive compound	Epigenetics modifications	Epigenetics regulators	Target	Mechanism	Biological function	Cancer type	References
Nigella sativa; family Ranunculaceae	India	Bibliographic review	Thymoquinone	Acetylation, Hypermethylation	DNMT1,3A,3B, G9A, HDAC1,4,9, KDM1B, KMT2A, B, C, D y E	DNMT1, G9a Y HDAC1	Induce apoptosis by inhibiting deacetylation	Anticancer activity	Breast cancer	Khan et al. (2019)
Radix angelicae sinensis	China	Experimental study	Z-Ligustilide	hypermethylation	HO-1, NQO1 y UGT1A1	CpG del Nrf2	Target gene re-expression: restoration of Nrf2	Anticancer activity	prostate cancer	Su et al. (2013)
Artemisia Sieversiana	Asia, Europe, and North America	Bibliographic review	Ethanolic extract of aerial parts		EIF2 y eIF4/ p70S6K	EIF2 y eIF4/ p70S6K	Induction of apoptosis, DNA damage, and loss of potential mitochondrial membrane	Cytotoxic activity	Cáncer colorrectal	Benarba and Pandiella (2018)
Withania somnifera	Asia and South Africa	Bibliographic review	5,6-de-Epoxy-5-en-7-one-17-hydroxy withaferin A	hypermethylation and histone modification	BIR5 protein	p53 y pRB	Prevent cancer cell formation and proliferation through gene silencing	Anticancer activity	prostate cancer	Rai et al. (2016)
Lycopodium clavatum	Spain	Bibliographic review	Apigenina		NF-KB	NF-KB	It eliminates cyclobutane rings and confers antiproliferative properties and cell apoptosis	DNA protection	Skin Cancer	George et al. (2017)

Vitis vinifera	Mediterranean Europe and Central Asia	Bibliographic review	Resveratrol	miRNA modifications	Downregulation miRNAs21 and 155 upregulation miR125 and 126	Pten-AKT-mTOR, FOX3, IGFB associaded to cell migration	Induces apoptosis	Micro RNA regulation	Breast cancer	Ahmed et al. (2020)
Camellia sinesis (green tea)	Japón, China e India	Bibliographic review	epigalocatequina-3-galato	miRNA-16 downregulation	miRNA-16	Bcl-2 in the HepG-2 cell line	Induces apoptosis	Micro RNA regulation	Hepatocellular carcinoma	Ahmed et al. (2020)
Ceratonia siliqua	Mediterranean	Bibliographic review	Pyrogallol, catechol, and phenolic acids	hypermethylation	Reversion of DNA hypermethylation	PARP, p53	Reversion of DNA hypermethylation	Anticancer activity	colon adenocarcinoma cells HCT-116 and CT-26	Benarba and Pandiella (2018)

One of its main bioactive components is bisdemethoxycurcumin, which can fight lung cancer cells due to its epigenetic mechanisms that could modify the Wnt/β-catenin signaling pathways. It could also negatively regulate the nuclear levels of β-catenin in β-catenin cancer cells. This could be done by inducing WIF1 re-expression (Echeverry et al. 2016).

Lycopodium clavatum

This is a species native to Spain, is used in various medical formulations to treat skin irritations and acne, nosebleeds, liver inflammation, kidney and bladder ailments, irritation of the intestinal tract and kidney disorders. Traditional medicine also uses it to heal wounds, relieve fatigue and muscle pain, induce pregnancy, as a carminative, nerve tonic, expectorant, and diuretic, and to treat lung disease, burns, headaches and childbirth disorders. This genus consists mainly of alkaloids, triterpenoids, glucosides, and to a lesser extent, monoterpenes, aliphatic alcohol, anthraquinone and two phytosterols. These components are responsible for their pharmacological properties such as inhibiting acetylcholinesterase, neuroprotective, anti-inflammatory, antimicrobial, antiviral, cardioprotective, antidiabetic, protection of hearing function, healing and antitumor (Wang et al. 2021).

One of its most predominant bioactive components for its activity against skin cancer is a flavone called apigenin, which protects the genetic material from UV-B radiation by inhibiting the development of ROS, the negative regulation of NF-KB and the elimination of cyclobutane rings. In addition, it may activate the phosphorylation of ATM and H2AX, thus regulating genes that control the cell cycle and DNA repair, which could subsequently lead to apoptosis. On the other hand, it has been reported that it could inhibit casein kinase two, regulating cell proliferation, mediating genetic damage and activation of NF-κB in cancer cells, and preventing DNA damage in normal cells (George et al. 2017).

Camellia sinensis

A medicinal species native to Japan, China, and India, commonly known as Miang, is of great medicinal and cultural importance and is used as a condiment. Among its main constituents are phenolic compounds such as flavanols, hydroxyl-4-flavonols, anthocyanins, flavones, flavonols, and phenolic acids, which provide it with antimicrobial, antioxidant, antimutagenic, anticancer, antitumor and antidiabetic effects. It also contains alkaloids of the methylxanthine type, which have beneficial effects on the nervous, cardiovascular, renal, respiratory and gastrointestinal systems (Khanongnuch et al. 2017).

One of its main bioactive components, epigallocatechin-3-gallate, significantly inhibits cancer cell viability and prevents tumor growth, which may inhibit the expression of miR-25 and apoptosis-related proteins. Another epigenetic effect of this molecule is to optimize the expression of 13 miRNAs, up-regulate miR-16 expression in HCC by cell death, reduce the expression of 48 miRNAs in HepG2 cancer cells, increase pro-caspase-9, pro-caspase-3 and PARP at the protein level and inhibit the expression of miR-2 (Ahmed et al. 2020).

REFERENCES

Adewumi, Odunayo Ayodeji, Vijender Singh, and Gunjan Singh. 2020. "Chemical Composition, Traditional Uses and Biological Activities of *Artemisia Species*." *Int J Pharmacogn Phytochem* 9 (5): 1124–1140.

Ahmed, Fayyaz, Bushra Ijaz, Zarnab Ahmad, Nadia Farooq, Muhammad Bilal Sarwar, and Tayyab Husnain. 2020. "Modification of MiRNA Expression through Plant Extracts and Compounds against Breast Cancer: Mechanism and Translational Significance." *Phytomedicine* 68: 153168. https://doi.org/https://doi.org/10.1016/j.phymed.2020.153168

Ayati, Zahra, Mahin Ramezani, Mohammad S Amiri, Ali Tafazoli Moghadam, Hoda Rahimi, Aref Abdollahzade, Amirhossein Sahebkar, and Seyed Ahmad Emami. 2019. "Ethnobotany, Phytochemistry and Traditional Uses of *Curcuma* Spp. and Pharmacological Profile of Two Important Species (*C. longa* and *C. zedoaria*): A Review." *Current Pharmaceutical Design* 25 (8): 871–935.

Azab, Abdullatif. 2017. "Carob (*Ceratonia siliqua*): Health, Medicine and Chemistry." *Eur Chem Bull* 6 (10): 456–469.

Benarba, Bachir, and Atanasio Pandiella. 2018. "Colorectal Cancer and Medicinal Plants: Principle Findings from Recent Studies." *Biomedicine & Pharmacotherapy* 107: 408–423. https://doi.org/https://doi.org/10.1016/j.biopha.2018.08.006

Echeverry, Andrés Hernán Cardona, Diego Fernando Uribe Yunda, and Fabian Mauricio Cortés-Mancera. 2016. "Actividad Antitumoral de La Curcumina Asociada a La Regulación de Mecanismos Epigenéticos: Implicaciones En La Vía Wnt/-Catenina." *Revista Cubana de Plantas Medicinales* 21 (4): 1–22.

Figueira, Leticia, and Julio César González. 2018. "Efecto Del Resveratrol Sobre Las Concentraciones Séricas Del Factor de Crecimiento Endotelial Vascular Durante La Aterosclerosis." *Clínica e Investigación En Arteriosclerosis* 30 (5): 209–216. https://doi.org/https://doi.org/10.1016/j.arteri.2018.04.003

George, Vazhappilly Cijo, Graham Dellaire, and H P Vasantha Rupasinghe. 2017. "Plant Flavonoids in Cancer Chemoprevention: Role in Genome Stability." *The Journal of Nutritional Biochemistry* 45: 1–14. https://doi.org/https://doi.org/10.1016/j.jnutbio.2016.11.007

Grandtner, M. M., and Julien B T—Dictionary of South American Trees Chevrette, eds. 2013. "Dictionary of South American Trees." In *Nomenclature, Taxonomy and Ecology Volume 2*, 55–86. San Diego: Academic Press. https://doi.org/https://doi.org/10.1016/B978-0-12-396490-8.00002-1

Griñán-Ferré, Christian, Aina Bellver-Sanchis, Vanessa Izquierdo, Rubén Corpas, Joan Roig-Soriano, Miguel Chillón, Cristina Andres-Lacueva, et al. 2021. "The Pleiotropic Neuroprotective Effects of Resveratrol in Cognitive Decline and Alzheimer's Disease Pathology: From Antioxidant to Epigenetic Therapy." *Ageing Research Reviews* 67: 101271. https://doi.org/https://doi.org/10.1016/j.arr.2021.101271

Khan, Md. Asaduzzaman, Mousumi Tania, and Junjiang Fu. 2019. "Epigenetic Role of Thymoquinone: Impact on Cellular Mechanism and Cancer Therapeutics." *Drug Discovery Today* 24 (12): 2315–2322. https://doi.org/https://doi.org/10.1016/j.drudis.2019.09.007

Khanongnuch, Chartchai, Kridsada Unban, Apinun Kanpiengjai, and Chalermpong Saenjum. 2017. "Recent Research Advances and Ethno-Botanical History of Miang, a Traditional Fermented Tea (*Camellia Sinensis* Var. Assamica) of Northern Thailand." *Journal of Ethnic Foods* 4 (3): 135–144. https://doi.org/https://doi.org/10.1016/j.jef.2017.08.006

Li, Xican, Xiaoting Wu, and Ling Huang. 2009. "Correlation between Antioxidant Activities and Phenolic Contents of *Radix Angelicae Sinensis* (Danggui)." *Molecules* . https://doi.org/10.3390/molecules14125349

Mahanna, Mohammed, Maria C Millan-Linares, Elena Grao-Cruces, Carmen Claro, Rocío Toscano, Noelia M Rodriguez-Martin, Maria C Naranjo, and Sergio Montserrat-de la Paz. 2019. "Resveratrol-Enriched Grape Seed Oil (*Vitis vinifera* L.) Protects from White Fat Dysfunction in Obese Mice." *Journal of Functional Foods* 62: 103546. https://doi.org/https://doi.org/10.1016/j.jff.2019.103546

Majeed, Abdul, Zahir Muhammad, Habib Ahmad, Rehmanullah, Sayed Sardar Sikandar Hayat, Naila Inayat, and Saira Siyyar. 2021. "*Nigella sativa* L.: Uses in Traditional and Contemporary Medicines—An Overview." *Acta Ecologica Sinica* 41 (4): 253–258. https://doi.org/https://doi.org/10.1016/j.chnaes.2020.02.001

Martínez, Arceo. 2018. "Modificaciones Epigenéticas de La Histona 3 Asociadas a La Actividad Citotóxica de La Defensina Ɣ-Tionina (*Capsicum chinense*) Sobre Células de Cáncer de Mama."

Mostafa, Tarek M, Sahar K Hegazy, Sherin S Elnaidany, Walid A Shehabeldin, and Eman S Sawan. 2021. "*Nigella sativa* as a Promising Intervention for Metabolic and Inflammatory Disorders in Obese Prediabetic Subjects: A Comparative Study of *Nigella sativa* versus Both Lifestyle Modification and Metformin." *Journal of Diabetes and Its Complications* 35 (7): 107947. https://doi.org/https://doi.org/10.1016/j.jdiacomp.2021.107947

Pellicer, Jaume, C. Haris Saslis-Lagoudakis, Esperança Carrió, Madeleine Ernst, Teresa Garnatje, Olwen M Grace, Airy Gras, et al. 2018. "A Phylogenetic Road Map to Antimalarial *Artemisia species*." *Journal of Ethnopharmacology* 225: 1–9. https://doi.org/https://doi.org/10.1016/j.jep.2018.06.030

Rai, Mahendra, Priti S Jogee, Gauravi Agarkar, and Carolina Alves dos Santos. 2016. "Anticancer Activities of *Withania somnifera*: Current Research, Formulations, and Future Perspectives." *Pharmaceutical Biology* 54 (2): 189–197. https://doi.org/10.3109/13880209.2015.1027778

Rayees, Sheikh, and Fayaz Malik. 2017. "*Withania somnifera*: From Traditional Use to Evidence Based Medicinal Prominence." *Science of Ashwagandha: Preventive and Therapeutic Potentials*, 81–103.

Su, Zheng-Yuan, Tin Oo Khor, Limin Shu, Jong Hun Lee, Constance Lay-Lay Saw, Tien-Yuan Wu, Ying Huang, et al. 2013. "Epigenetic Reactivation of Nrf2 in *Murine prostate* Cancer TRAMP C1 Cells by Natural Phytochemicals Z-Ligustilide and *Radix Angelica Sinensis* via Promoter CpG Demethylation." *Chemical Research in Toxicology* 26 (3): 477–485. https://doi.org/10.1021/tx300524p

Velázquez-Hernández, María Elena, Alejandra Ochoa-Zarzosa, and Joel E López-Meza. 2021. "Defensin γ-Thionin from *Capsicum chinense* Improves Butyrate Cytotoxicity on Human Colon Adenocarcinoma Cell Line Caco-2." *Electronic Journal of Biotechnology* 52: 76–84. https://doi.org/https://doi.org/10.1016/j.ejbt.2021.04.009.

Wang, Bo, Canyuan Guan, and Qiang Fu. 2021. "The Traditional Uses, Secondary Metabolites, and Pharmacology of *Lycopodium* Species." *Phytochemistry Reviews*. https://doi.org/10.1007/s11101-021-09746-4

Zheng, Tao, and Hainan Chen. 2021. "Resveratrol Ameliorates the Glucose Uptake and Lipid Metabolism in Gestational Diabetes Mellitus Mice and Insulin-Resistant Adipocytes via MiR-23a-3p/NOV Axis." *Molecular Immunology* 137: 163–173. https://doi.org/https://doi.org/10.1016/j.molimm.2021.06.011

8 *Bursera graveolens* (Palo Santo), Native Species of the Ecuadorian Dry Forest

Patricia Manzano, Andrea Orellana-Manzano and Glenda Pilozo*

CONTENTS

INTRODUCTION BOTANICAL DESCRIPTION

B. graveolens, commonly referred to as Palo santo, according to the Dictionary of South American Trees may also be referred to as *Amyris caranifera, Bursera penicillata, Bursera pubescens, Bursera tatamaco, Elaphrium graveolens, Elaphrium penicillatum, Elaphrium pubescens, Elaphrium tatamaco, Spondias edmonstonei, Terebinthus graveolens and Terebinthus pubescens;* it belongs to the Bruceraceae family, native to the coast of Mexico, Colombia, Ecuador, Peru and Brazil (Grandtner et al., 2013).

It is a resinous and aromatic tree, 3 to 15 m long, the stem reaching up to 40 cm of diameter, whose bark is gray, with smooth covering; its leaves are 30 cm long and 18 cm broad, with obtuse or pointed apex and irregular margin; the leaflets are 3 to 9 cm long by 1 to 4 cm wide, alternately distributed at the end of the branches; the petiole is up to 9 cm long, rachis with narrow wings; their male tetramer flowers, turbinated calyx, their tiny triangular lobes, less than 0.5 mm long, petals whitish, yellowish or greenish in the shape of an elliptical shape; female flowers have a similar structure to the male, stamina with anthers less than 0.6 mm long, bilocular ovary, very short style, two stigmata; bivalve fruit, sub-spherical to obovoide, 6–10 mm long, yellowish, white or green; its fruits are smooth up to 1 cm; the seeds are black covered by an orange-red bark; the wood has a strong sweet and persistent odor, characteristic of

DOI: 10.1201/9781003173991-11

the species (Jaramillo-Colorado et al., 2019; Fon-Fay et al., 2019; Marcotullio et al. 2018; Paniagua-Zambrana et al., 2020).

GENETICAL AND PHENOLOGICAL CHARACTERISTICS

Through a molecular phylogenetic analysis of various species of *Bursera*, endemic to the Galapagos Islands, evidence of a common ancestor and a possible lineage of miscegenation was found. However, despite the lack of genetic diversity shown by population genetic analysis of AFLP data, a slight morphological difference between species is also observed. Phenotypic identification is of vital importance for the differentiation of species and subspecies (Weeks et al., 2009).

FIGURE 8.1 Vegetative development stage of *Bursera graveolens*.

FIGURE 8.2 Fruiting stage of *Bursera graveolens*.

The phenological events of *Bursera graveolens* vary according to the climatic and geographical conditions of its location. For the species located to the south of Ecuador, the flowering occurs in the month of January, the fructification begins in the month of February and culminates in the month of August, with greater intensity in the first 3 months of this phase; defoliation begins in April, remaining constant until December (Infante et al., 2016). In Northern Peru, flowering begins in January and stops in July, the process restarts in the last two months of the year; fruiting begins in January and ends in June; defoliation remains from the beginning of July until the end of October (Martos et al., 2009).

TRADITIONAL USES

This species has been used since ancient times for religious, commercial and ethnobotanical purposes (DeCarlo et al., 2019) as an analgesic, anti-inflammatory, antitumor, antidiarrheal, expectorant, depurative, diaphoretic, healing, abortive, analgesic, insecticide, insect repellent, for the treatment of anemia, rheumatism, dermatitis, asthma, colic and rheumatic pains (Jaramillo-Colorado, 2019; Grandtner et al., 2013).

Locally in Ecuador, the dry bark is used as a fragrance and to treat anemia; the powdered stems are used to calm the stomach. In Colombia, exudate resin is used to treat hernias, sprains and to remove foreign bodies from the skin; the trunk and fruits are consumed to treat arthritis, muscle pain, skin allergies, asthma, cough, headache, stress, flu, dizziness, as a circulatory stimulant; the bark is depurative, diuretic and diaphoretic (Paniagua-Zambrana et al., 2020).

Of the essential oil, its aromatic properties are highlighted by the high content of volatile compounds, attributing to it the properties of anti-inflammatory, fungicidal, insecticidal and bactericidal; having a repellent effect of 88.1% against *T. castaneum* (Jaramillo-Colorado, 2019); an acaricide effect of the essential oil against larvae of *R. microplus* (Rey-Valeirón et al. 2017, 344); analgesic, antimicrobial, antiviral, anti-tumor, anticonvulsant effects have also been reported (Fon-Fay et al., 2019).

The potential use of waste generated by the extraction of essential oil in the production of enzymes by biodegradation with specific fungi has also been recognized from an industrial perspective (Carrión-Paladines et al., 2019).

MAIN CONSTITUENTS

The qualitative determination of the active ingredients present in the branches of *B. graveolens* records the content of triterpenes, steroids, phenols, tannins, flavonoids, saponins and amino acids. Extraction with methanol was used to isolate lignans: burseranin, picropoligamain and triterpenes: lupeol and epi-lupeol, compounds with high biological interest (Marcotullio et al., 2018).

The following compounds were isolated from the essential oil of the branches of *B. graveolens*: D-Limonene, 2-Cycloheexen-1-ol,2-methyl-5-(1-methylethenyl), fenchona, Viridiflorol, 1,2-Cyclohexanediol,1-methyl-4-(1-methylethienyl), 3,4-methylenedioxyacetophenone, Spatulenol, 6-isopropenyl-4,8α-dimethyl-1,2,3,6,7,8,8α-octahydronaphthalen-2-ol, α-Copaeno, Naphthalene-1,2,3,5,6,8α-hexahydro-4–7-dimethyl-1, α-Cadinol (Manzano et al., 2009).

However, it is now known that the chemical constitution of the species is correlated with the geographical location of the origin and the method of extraction. By means of simultaneous distillation-extraction, the content of limonene (9.1%), α-terpineol (8.1%) and β-bisabolene (5.7%) has been reported in the species originating in Peru (Yukawa et al. 2006, 234); mint furanone (43.9%), iso-mint furanone (6.8%) and 3-hydroxy- mint furanone (6.2%) in the species native to the west coast of Colombia (Muñoz-Acevedo et al. 2013, 322). The oil obtained by maceration at 60° in Western Ecuador shows 85% of D-Limonene and 14.7% mentofuran (Almeida et al., 2019).

Also, variability in the chemical constitution of oils obtained by traditional hydrodistillation has been reported, obtaining as a majority compound the monoterpene limonene hydrocarbon of species from Northern Colombia with 42.1% relative abundance (Leyva et al., 2007), the Midwest of Ecuador with 58.6% (Young et al., 2007) and 26.5% in Cuba; information that contrasts with the reported content for the species of Western Ecuador indicating as the majority component viridiflorol with 70.8% (Manzano et al., 2009). This variability corresponds to genetic, environmental factors and the processing techniques used to obtain the essential oil (Fon-Fay et al., 2019).

On the other hand, a multivariate analysis determined the major compounds of each organ of the species. On the leaves, germacre D, trans-β -caryophyllene, viridiflorol, limonene, linalool, dendrolasin; in immature fruit, 3-hydroxy- mint furanone, mint furanone, carvone, limonene, trans-carveol, limonen-1,2-diol, limonene dioxide derivative, 2-hydroperoxide-2S,4R-p-mentha-6,8-diene, and caryophyllene oxide; Bark, mint furanone, 3-hydroxy-mint furanone, iso-mint furanone; Branches, mint furanone, iso-mint furanone, 3-hydroxy-mint furanone; Resin, limonene, mint furanone, mint furanone derivative, pulegone, 3-hydroxy-mint furanone and mentofuran (Muñoz-Acebedo et al., 2013).

PHARMACOLOGICAL ACTIVITY

The antioxidant capacity of the essential oil has been demonstrated as it stabilizes diphenyl—2,4,6-trinitrophenyl; iminoazanium with an IC_{50}: 22.9 mg/mL and 2,4,6-tripiridyl-s-triazin with 354.0 μM ascorbic acid equivalent (Fon-Fay et al., 2019). In addition, antifungal activity was found with a minimum inhibitory concentration of 1.33 mg/mL against strains of *Candida utilis*; antibacterial property against *Staphylococcus aureus*, in the form of β-cyclodextrin microencapsulated inhibited the growth of ATCC 11229 (Calvo-Irabien, 2018).

Another study confirmed antibacterial activity of a medium to high range in the essential oil against strains of *E. coli, Bacillus cereus, Listeria monocytogenes, Clostridium perfringens, Salmonella choleraesuis* and *Candida albicans* (Mendez et al., 2017) and spasmolytic effects with the hexane extract of the leaves of the species that also inhibited the bacterial growth of the strains of Escherichia coli, *Salmonella enterica, Salmonella enteritidis* and *Salmonella choleraesuis*. It also exhibits antimalarial activity *in vitro* and *in vivo*, due to its antiproliferative capacity of *L. amazonensis* mixers with IC_{50}: 48 μg/mL (Monzote et al., 2012).

An *in vivo* study demonstrated the anti-inflammatory capacity of hydroalcoholic extracts of *B. graveolens* branches, with the capacity to reduce inflammation by 88.23%, comparable results with the effect shown by (Manzano et al., 2009). Its anti-inflammatory properties and the anticancer potential of lupeol and epi-lupeol, whose inhibition of human HT1080 fibrosarcoma is comparable to the reference drug, were also evaluated. Picropoligamain, isolated from the resin, inhibits the adenocarcinoma of the human prostate (Marcotullio et al., 2018) and the essential oil of the cortex has a cytotoxic effect on MCF-7 breast tumor cells (Nieto-Yañez et al., 2017).

On the other hand, the potential of active metabolites with therapeutic applications against SARS-CoV-2 of the compounds isolated from the leaves of *B. grandifolia* is reported (Sharma et al., 2020), a very relevant report to support this type of study with other species of the Bruceraceae family, specifically in *Bursera graveolens,* to thus increase its medicinal benefits and contribute to the health of the world population.

CONCLUSION

This species is a potential source of bioactive compounds with important pharmacological activities, whose concentration depends on the climatic and geographical conditions of its location, for which pharmaceutical, cosmetic and industrial products could be developed, however it is necessary to know in depth the safety of its use application.

REFERENCES

Almeida, Luis, and Loor David Coello Cedeño. 2019. "Estudio Comparativo de La Composición Química Del Aceite Medicinal de Palo Santo de Illari vs La Composición Química Del Aceite Esencial de Palo Santo." *Ciencia* 37: 223–224. 10.13140/RG.2.2.14858.82884

Calvo-Irabien, Luz María. 2018. "Native Mexican Aromatic Flora and Essential Oils: Current Research Status, Gaps in Knowledge and Agro-Industrial Potential." *Industrial Crops and Products* 111: 807–822. https://doi.org/https://doi.org/10.1016/j.indcrop.2017.11.044

Carrión-Paladines, Vinicio, Andreas Fries, Rosa E Caballero, Pablo Pérez Daniëls, and Roberto García-Ruiz. 2019. "Biodegradation of Residues from the Palo Santo (*Bursera graveolens*) Essential Oil Extraction and Their Potential for Enzyme Production Using Native Xylaria Fungi from Southern Ecuador." *Fermentation* . 5(76): 1–20. https://doi.org/10.3390/fermentation5030076.

DeCarlo, Anjanette, Noura S. Dosoky, Prabodh Satyal, Aaron Sorensen, and William N. Setzer. 2019. "The Essential Oils of the Burseraceae." Essential Oil Research. *Springer, Cham*. 61–145. https://doi.org/https://doi.org/10.1007/978-3-030-16546-8_4.

Fon-Fay, Flor M., Jorge A. Pino, Ivones Hernández, Idania Rodeiro, and Miguel D. Fernández. 2019. "Chemical Composition and Antioxidant Activity of *Bursera graveolens* (Kunth) Triana et Planch Essential Oil from Manabí, Ecuador." *Journal of Essential Oil Research* 31(3): 211–216. https://doi.org/10.1080/10412905.2018.1564381

Grandtner, M M, and Julien B T—Dictionary of South American Trees Chevrette, eds. 2013. "Dictionary of South American Trees." *In Nomenclature, Taxonomy and Ecology* 2: 55–86. *San Diego: Academic Press*. https://doi.org/https://doi.org/10.1016/B978-0-12-396490-8.00002-1

Infante, Luis Felipe Morillo, Víctor Hugo Eras Guamán, José Moreno Serrano, Julia Minchala, Patiño, Luis Muñoz Chamba, Magaly Yaguana Arévalo, Ruth Poma Angamarca, Cristian Valarezo, and Ortega. 2016. "Phenological Study and Propagation of *Bursera graveolens* (Kunth) Triana & Planch, in the Community of Malvas, Canton Zapotillo, Province of Loja." *Bosques Latitud Cero* 6(2): 41–63.

Jaramillo-Colorado, Beatriz Eugenia. Samyr Suarez-López. Vanessa Marrugo-Santander. 2019. "Volatile Chemical Composition of Essential Oil from *Bursera graveolens* (Kunth) Triana & Planch and Their Fumigant and Repellent Activities." *Biological Sciences* 41(1): e46822. https://doi.org/10.4025/actascibiolsci.v41i1.46822

Joaquín, Martos, Mariella Scarpati, Consuelo Rojas y Guillermo E. Delgado. 2009. "Phenology of Some Species That Are Food for the White-Winged Guan Penelope Albipennis." *Rev. Peru. Biol* 15(4): 51–58. https://doi.org/10.15381/rpb.v15i2.1721

Leyva, M., Martínez, J., and Stashenko, E. 2007. "Composición Química Del Aceite Esencial de Hojas y Tallos de *Bursera graveolens* (Burseraceae) de Colombia." *Scientia Et Technica* 1(33): 201–202. https://doi.org/10.22517/23447214.6099

Manzano Santana, P., M. Miranda, Y. Gutiérrez, G. García, T. Orellana, and A. Orellana. 2009. "Antiinflammatory effect and chemical composition of *Bursera graveolens* Triana & Planc. branch oil (Palo Santo) from Ecuador." *Revista Cubana de Plantas Medicinales* 14(3): 45–53.

Marcotullio, Maria Carla, Massimo Curini, and Judith X Becerra. 2018. "An Ethnopharmacological, Phytochemical and Pharmacological Review on Lignans from Mexican *Bursera Spp*." *Molecules* (Basel, Switzerland) 23(8): 1–6. https://doi.org/10.3390/molecules23081976

Mendez, Alejandrina Honorata Sotelo, Clara Gabina Figueroa Cornejo, Mary Flor Césare Coral, and M Arnedo. 2017. "Chemical Composition, Antimicrobial and Antioxidant Activities of the Essential Oil of *Bursera graveolens* (Burseraceae) from Perú." *Indian Journal of Pharmaceutical Education and Research* 51(3): S429–S435. https://doi.org/10.5530/ijper.51.3s.62

Monzote, Lianet, Gabrielle M Hill, Armando Cuellar, Ramón Scull, and William N Setzer. 2012. "Chemical Composition and Anti-Proliferative Properties of *Bursera graveolens* Essential Oil." *Natural Product Communications* 7(11): 1531–1534. https://doi.org/10.1177/1934578X1200701130

Muñoz-Acebedo, Amner, Serrano-Uribe, Adriana, Parra-Navas, Ximena J., Olivaresescobar, Luz Andrea, Niño-Porras, Mónica E. 2013. "Análisis Multivariable y Variabilidad Química de Los Metabolitos Volátiles Presentes En Las Partes Aéreas y La Resina de *Bursera graveolens* (Kunth) Triana & Planch . de Soledad (Atlántico, Colombia)." *Bol Latinoam Caribe Plant Med Aromat* 12(3): 322–337.

Muñoz-Acevedo A., A. Serrano-Uribe, X.J. Parra-Navas Niño-Porras, L.A. Olivares-Escobar and M.E. 2013. "Multivariate Analysis and Chemical Variability of the Volatile Metabolites Present in the Aerial Parts and the Resin of *Bursera graveolens* (Kunth) Triana & Planch from Soledad (Atlántico, Colombia)." *Boletin Latinoamericano y Del Caribe de Plantas Medicinales y Aromaticas* 12(3): 322–337.

Nieto-Yañez, O. J., A. A. Resendiz-Albor, P. A. Ruiz-Hurtado, N Rivera-Yañez, M. Rodriguez-Canales, M. Rodriguez-Sosa, I. Juarez-Avelar, M. G. Rodriguez-Lopez, M. M. Canales-Martinez, and M. A. Rodriguez-Monroy. 2017. "*In Vivo* and *in Vitro* Antileishmanial Effects of Methanolic Extract from Bark of *Bursera aptera*." *African Journal of Traditional, Complementary, and Alternative Medicines* : AJTCAM 14(2): 188–197. https://doi.org/10.21010/ajtcam.v14i2.20

Paniagua-Zambrana, Narel Y., Rainer W. Bussmann, and Carolina Romero. 2020. "*Bursera graveolens* (Kunth.) Triana & Planch. Burseraceae." *Ethnobotany of the Andes*, 1–5. https://doi.org/10.15468/39omei.

Rey-Valeirón, Catalina, Lucía Guzmán, Luis Rodrigo Saa, Javier López-Vargas, and Eduardo Valarezo. 2017. "Acaricidal Activity of Essential Oils of *Bursera graveolens* (Kunth) Triana & Planch and Schinus Molle L. on Unengorged Larvae of Cattle Tick Rhipicephalus (Boophilus) Microplus (Acari:Ixodidae)." *Journal of Essential Oil Research* 29(4): 344–50. https://doi.org/10.1080/10412905.2016.1278405

Sharma, Namisha, Mehanathan Muthamilarasan, Ashish Prasad, and Manoj Prasad. 2020. "Genomics Approaches to Synthesize Plant-Based Biomolecules for Therapeutic Applications to Combat SARS-CoV-2." *Genomics* 112(6): 4322–4331. https://doi.org/10.1016/j.ygeno.2020.07.033

Weeks, Andrea, and Alan Tye. 2009. "Phylogeography of Palo Santo Trees (*Bursera graveolens* and *Bursera malacophylla*; Burseraceae) in the Galápagos Archipelago." *Botanical Journal of the Linnean Society* 161(4): 396–410. https://doi.org/10.1111/j.1095-8339.2009.01008.x

Young, D. G., S. Chao, H. Casabianca, M.-C. Bertrand, and and D. Minga. 2007. "Essential Oil of *Bursera graveolens* (Kunth) Triana et Planch from Ecuador." *Journal of Essential Oil Research* (19): 525–526. https://doi.org/10.1080/10412905.2007.9699322

Yukawa, C., Y. Imayoshi, H. Iwabuchi And, Komemushi, and A. Sawabe. 2006. "Chemical Composition of Three Extracts of *Bursera graveolens*." *Flavour and Fragrance Journal* 21: 234–238. https://doi.org/10.1002/ffj.1563

9 Pharmacognostic, Chemical and Pharmacological Studies of *Corynaea crassa* Growing in Ecuador

Alexandra Jenny López Barrera

CONTENTS

DOI: 10.1201/9781003173991-12

INTRODUCTION

Medicinal plants have a very important contribution to the health system at a local and global level because they are frequently used by most rural populations (Angulo et al., 2012). Currently, a large percentage of the world's population, particularly in developing countries, use plants to meet primary health care needs (Tene et al., 2007).

Ecuador is within the group of countries enriched in flora and fauna, representing more than 70% of biodiversity worldwide, and due to its high biological and cultural variety, it constitutes one of the countries with great potential in terms of traditional medicine (Zambrano et al., 2015). The Ecuadorian flora is supplied with a great versatility of plant species for medicinal use, but there is a percentage of species with possible phytotherapeutic properties that have little documented information (Armijos et al., 2014).

In Ecuador, studies of medicinal plants and ethnobotany have been developed, mainly in the Andean region (Armijos et al., 2014), where it is part of the daily practices of its inhabitants due to vast ancestral medical knowledge. It is reported that 3,118 species belonging to 206 plant families are used for medicinal purposes, within which the families that present a greater number of species are Asteraceae, Fabaceae, Rubiaceae, Solanaceae and Aracaceae (Caballero-Serrano et al., 2019).

In the neighboring country of Peru, a wide variety of medicinal plants are used in contemporary popular medicine for aphrodisiac purposes, to modulate fertility or to induce sterility (Malca et al., 2015). One of the plants used in these aphrodisiac mixtures is the species *Corynaea crassa*, commonly known as Chutarpo or Huanarpo, with the masculine designation "Huanarpo macho" or "Peruvian viagra" (Bussmann, 2010), which is drunk as an alcoholic beverage. There are also reports of ethanolic extracts that have shown antimicrobial activity against *Staphylococcus aureus* (Bussmann, 2006; Bussmann et al., 2011). However, currently, there is little information on its chemical components (Malca et al., 2015).

The species *Corynaea crassa* Hook. F. (Balanophoraceae) is a hemiparasitic plant native to America, found in various countries, including: Bolivia, Colombia, Costa Rica, Ecuador, Mexico, Panama, Peru and Venezuela. In Ecuador it is found in the provinces of Azuay, Carchi, Chimborazo, Cotopaxi, Imbabura, Loja, Morona Santiago, Napo, Pichincha, Sucumbios, Tungurahua and Zamora Chinchipe (Tropicos, 2019). In Peru, this plant species is widely used in traditional medicine for the treatment of erectile dysfunction (Malca et al., 2015), but of which there is no scientific information that supports this popular knowledge. This plant also grows in the Andean regions of Ecuador, but it is not widely known among the population. For this reason, it was proposed to carry out pharmacognosy, phytochemistry and biological studies in the plant that allows to provide updated information about it so this natural resource can be used for the benefit of humankind.

The species *C. crassa* Hook. F., is widely used in traditional Peruvian medicine for the treatment of erectile dysfunction and has been shown to have antimicrobial properties. However, there are few phytochemical studies and there is a lack of research from the genetic, pharmacognostic and biological point of view, which would allow establishing its quality, efficacy and safety as a plant medicine.

CORYNAEA CRASSA HOOK. F.

DEFINITION

Corynaea crassa (Figure 9.1) is a hemiparasitic plant which grows on roots of other species. It belongs to the Balanophoraceae family (Brako and Zarucchi, 1993; Lobo, 2003; Tupac et al., 2009; Tropicos, 2019) and has been recorded on a variety of host plants (Catalogue of the Flowering plants, 2014).

SYNONYMS

Corynaea purdiei Hook. F, *Corynaea sphaerica* Hook. F., *Corynaea sprucei* Eichler, *Itoasia crassa* Hook. F., *Kuntze Itoasia* purdiaei Hook. F. Kuntze, *Itoasia sphaerica* Hook. F. (Kuntze-Hansen, 1993; Catalogue of the Flowering plants, 2014; Sato and Gonzalez, 2013).

SELECTED VERNACULAR NAMES

Corynaea crassa is under the male designation "Huanarpo male" and is conventionally used as an aphrodisiac for men (Hansen, 1983; Pachacuti-Yamqui, 1992; Bussmann, 2006; Bussmann, 2010; Sato and Gonzalez, 2013; Catalogue of the Flowering plants, 2014). The common name is Mazorquita (Catalogue of the Flowering plants, 2014).

FIGURE 9.1 *Corynaea crassa* Hook. F. in natural habitat. Photographs taken by the authors during the collection of the plant.

Description

They are brownish yellow deep purple plants that become brown or blackish when dry. Its roots measure about 3–6 × 4–8 cm in length. Stems usually measure 3–6 × 0.5–2 cm, which emerge directly from the tuber, breaking the surface on emergence and forming a conspicuous ring at the base of each stem. These are slightly longitudinally ribbed in the distal half when dry. Inflorescences 3–7 × 2–5 cm without the bracts, unbranched; bracts 4–8 mm in diameter, flattened or with a conical protuberance in the center; filiform trichomes 1.8–2 mm, claviform, pink. Flowers staminated with perianth tube 3.5–5 mm, wolves 1.8 × 0.8 mm, ligulate; stamens with column 2.5–6 mm, attached to the perianth tube in the proximal 1/3; sinandro c. 1 × 1 mm. Flowers pistillate with the segments of the perianth 1–1.2 mm, protruding from the ovary; styles 1–1.2 mm, exerted from filiform trichomes (Gómez, 1983; Knapp, 2012).

GEOGRAPHICAL DISTRIBUTION

It is native to America, which is located in the high Andean forests, with an altitude range of 1,300 to 3,000 meters above the sea level (Bussmann, 2006). Here we report observations of neotropical species distributed from Costa Rica to all the Andean countries except Chile (Tropicos, 2007).

In Ecuador, harvesting occurs in tropical forests of Azuay, Carchi, Chimborazo, Cotopaxi, Imbabura, Loja, Morona Santiago, Napo, Pichincha, Sucumbíos, Tungurahua and Zamora Chinchipe at an altitude of approximately 1,050 to 3,000 meters above the sea level (Tropicos, 2019).

PLANT MATERIAL OF INTEREST

Macroscopic inspection and micromorphological characteristics (Miranda and Cuellar, 2000), microscopic characteristics (Peacock, 1973; Gattuso and Gattuso, 1999; Catalogue of the Flowering plants, 2014) and physical chemical parameters of the raw drug, phytochemical screening for the identification of secondary metabolites (Miranda and Cuellar, 2000).

ORGANOLEPTIC PROPERTIES

They are brownish yellow to deep purple, brown or blackish when dry (Catalogue of the Flowering plants, 2014).

MACROMORPHOLOGICAL EVALUATION

It was found that the plant is reddish brown or brown (a). The upper end has a rounded to ovoid shape, with a spongy-fibrous consistency (b and c), followed by an elongated piece slightly ribbed on its external surface (d) and ribbed fibrous (e) on its internal surface. The lower part shows the haustorial root (Ff), protruding, rough, which, when dried, fragments leave fibers. Figure 9.2 shows some morphological details of

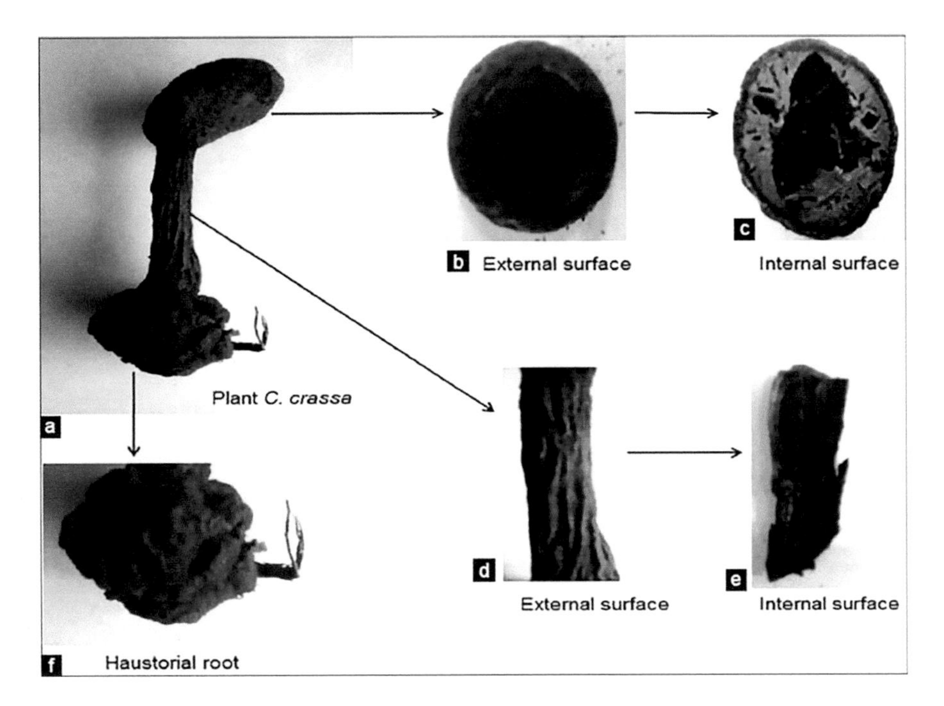

FIGURE 9.2 Representative Macromorphological details from *C. crassa*. Photographs taken by authors during pharmacognostic analysis of samples from Ecuador. (a) Plant *Corynaea crassa*, (b) External surface, (c) Internal surface, (d) External surface, (e) Internal surface, (f) Haustorial root.

the plant under study. Measurements of plant length were made, showing dimensions of 11.75 ± 2.10 cm (Lopez et al., 2021).

MICROMORPHOLOGICAL EVALUATION

The analysis of the powdered drug, Figure 9.3 allowed us to observe xylem vessels of the ladder type (A). It was possible to visualize epidermal cells with a sinusoidal contour, together with other structures that respond to the name of stomata (C), which are responsible for the gaseous exchange of the plant with the external environment. Cells of the fundamental parenchyma (E) were also observed with small starch granules, which suggests that it could be an amyliferous reserve parenchyma (accumulates starches), in addition scleroid cells (G) that are part of the protection fabric (Lopez et al., 2021).

The samples were subjected to some histochemical reactions (Figure 9.3). Oil bags (A and B) could be seen with great clarity, thanks to the tests carried out with Sudan III, a reagent that dyes red oils or fatty compounds in different shades. Starch grains (C and D) were observed, which showed an intense violet-blue color compared to the Lugol reagent. The starch grains are of the lenticular type (Lopez et al., 2021).

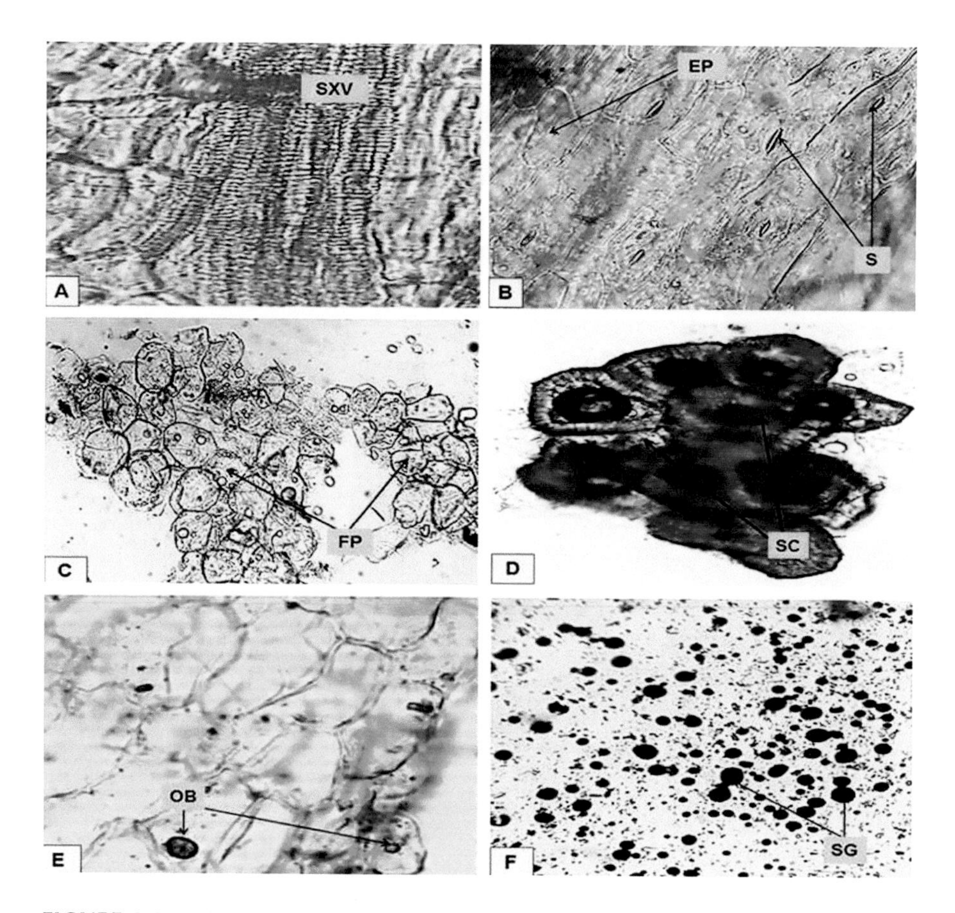

FIGURE 9.3 Micromorphological details of the powdered drugs *Corynaea crassa* from Ecuador. SXV: scaleriform xylem vessels (A); EP: epidermal cells (B), S: stomata (B); FP: fundamental parenchyma (C); SC: sclera cells (D); Test Sudan (E), OB; Test Lugol (F), SG; starch grains. Photographs were taken by the authors with an optical microscope at 10X

PHARMACOGNOSTIC PARAMETERS IN POWDERED PLANT MATERIAL

The different tests were carried out on raw drugs according to the methodology (Miranda and Cuellar, 2000) (Norma Ramal NRSP 309, 1992).

Physicochemical Parameters of the Crude Drug

- ***Residual humidity***: between 8–14% (Gómez, 1983, Miranda and Cuellar, 2000). It was reported: 8.64% ± 0.03 (Lopez et al., 2021).
- **Total ash:** no more than 15.0% (15). It was reported: 7.81 % ± 0.09 (Lopez et al., 2021).

- **Water soluble ash:** no more than 5% (19). It was reported *in vitro* 0.69 ± 0.02% (Lopez et al., 2021).
- **Alcohol-soluble extractive at 50%:** 29.68% ±0.18 (Lopez et al., 2021).
- **Alcohol-soluble extractive at 80%:** 44.76 % ± 0.11 (Lopez et al., 2021)
- **Alcohol-soluble extractive at 98%:** 13.69 % ± 0.24 (Lopez et al., 2021)
- **Ethyl acetate-soluble extractive:** 2.14 % ± 0.04 (Lopez et al., 2021)
- **Hexane-soluble extractive**: 2.00 % ± 0.04 (Lopez et al., 2021)
- **Acid-Insoluble ash:** to be established in accordance with national requirements. It was reported 5.37 ± 0.02 % (Lopez et al., 2021).

IDENTIFICATION OF SECONDARY METABOLITES BY PHYTOCHEMICAL SCREENING

Crude drug analysis was performed according to procedure (Gómez, 1983, Miranda and Cuellar, 2000).

Phytochemical screening of the crude drug revealed the presence of phenolic compounds, reducing substances, triterpenoids and steroids, as well as abundant saponins in the aqueous extract.

Physical-chemical parameters of the extracts (using an 80% hydroalcoholic mixture and water as solvents)

Hydroalcoholic extract

- ***pH:*** 4.75 ± 0.00
- ***Total solids:***14.43 ±0.26
- ***Refraction index:***1.3863 ±0.0002
- ***Relative density:***0.9760 ±0.0001

Aqueous extract

- ***pH***: 4.51 ± 0.02
- ***Total solids***: 1.67 ±0.01
- ***Refraction index***: 1.3306 ± 0.0001
- ***Relative density:*** 1.0166 ± 0.0008

CHEMICAL TESTS OF EXTRACTS OF *CORYNAEA CRASSA*

Chemical tests were conducted to assess total phenols by the Folin-Ciocalteu method (Ditjen. Materia medika Indonesia, 1995; Pourmorad et al., 2006) and total flavonoids by the colorimetric method of aluminum chloride (Chlopicka et al., 2012; Chang et al., 2002).

Through Gas chromatographic-Mass spectrometry (GC-MS) the allocation of the structures was made by comparing the sample mass spectra with the spectra from the team's libraries, Wiley 9th edition and NIST 2011, choosing those compounds that presented more than 95% reliability from the Center for Biotechnological Research of Ecuador CIBE—Espol.

TOTAL PHENOLS AND TOTAL FLAVONOIDS

- **Hydroalcoholic extract**
  - Total phenols: 43, 52 ± 0, 10 mg/mL
  - Total flavonoids: 5, 81 ± 0, 07 mg/mL
- **Aqueous extract**
  - Total phenols: 14, 90 ± 0, 12 mg/mL
  - Total flavonoids: 2, 37 ± 0.07mg/mL

Gas Chromatographic—Mass Spectrometry (GC-MS)

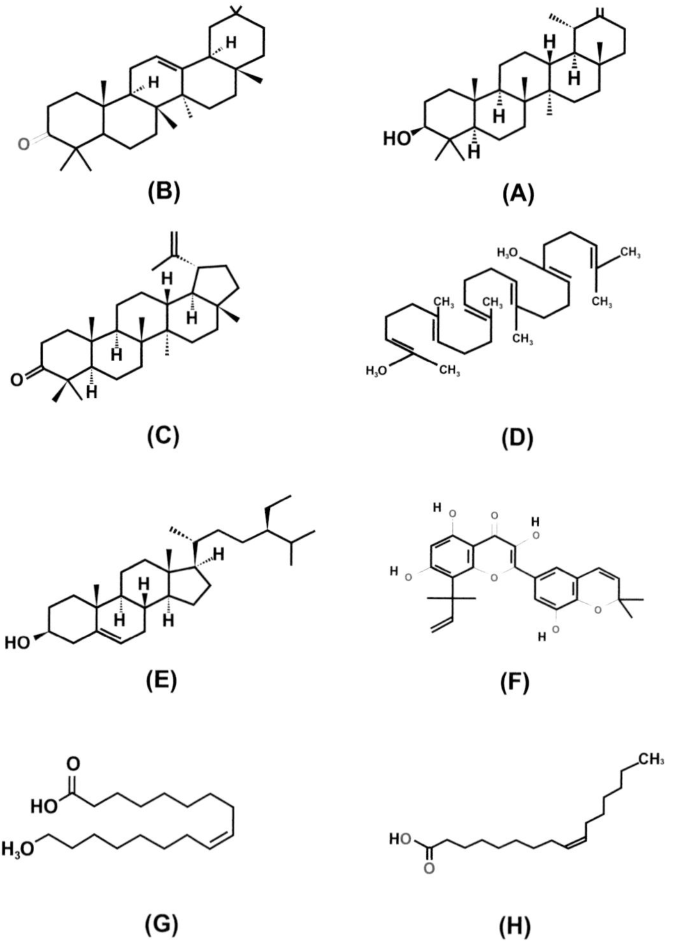

FIGURE 9.4 Major Components of *Corynaea crassa*. A) β-amyrin, (B) Taraxasterol, (C) Lupenone, (D) Squalene, (E) Sitosterol, (F) Broussonol A, (G) Octadecanoic acid, (H) Hexadecanoic acid.

MAJOR CHEMICAL CONSTITUENTS

The most important chemical constituents reported from 80% hydroalcoholic and aqueous extracts of *Corynaea crassa* were prepared by percolation and decoction respectively.

The aqueous extracts were rich in phenolic compounds such as pinorresinol, as well as sugars and fatty acids composed as lupenone, *β*-sitosterol, *β*-amyrone and *β*-amyrin.

The hydroalcoholic extracts were rich in sterol (sitosterol), terpenoids and fatty acid esters (Figure 9.4).

MEDICINAL USES

C. crassa is not only an aphrodisiac plant. There are also reports from ethanolic extracts which have shown biological activity against *Staphylococcus aureus* (Malca et al., 2015). The ethanolic extract shows a toxicity of 116 µg/mL using a brine-shrimp lethality test. However, the aqueous extract was found to be non-toxic (<10,000 µg/mL) for the same lethality test (Bussmann et al., 2011).

USES DESCRIBED IN FOLK MEDICINE, NOT SUPPORTED BY EXPERIMENTAL OR CLINICAL DATA

Uses described in Peruvian folk medicine, not supported by clinical data, include as an aphrodisiac, prepared as an alcoholic beverage known as "Huanarpo macho" (Malca et al., 2015; Fernandez et al., 2017; Winter et al., 1962; Gonzalez et al., 2011).

PHARMACOLOGY

Anti-Inflammatory Activity

Anti-inflammatory activity has been measured used the plantar edema test by carrageenan (Udupa et al., 1994). The anti-inflammatory activity of extracts of *Corynaea crassa* with a dose of 400 mg/kg by a number of *in vitro* studies has revealed that it reduces acute inflammation in rats (carrageenan-induced paw o edema), nonetheless no effect on chronic inflammation was observed. The anti-inflammatory effect was verified using the model of plantar edema in rats, induced by carrageenan applied subcutaneously within the surface of the aponeurosis of the right foot (Guo et al., 2019; Kosteret al., 1959).

The group treated with indomethacin showed the highest percentages of inhibition, mainly at 3 and 5 hours. The aqueous and hydroalcoholic extracts of *C. crassa*, although they had a lower percentage, also managed to significantly reduce the edema induced by carrageenan in the paw of the rat, from the third and fifth hour, showing evidence as anti-inflammatory under the conditions tested.

Analgesic Effects: Acetic Acid-Induced Abdominal Contortions in Mice

Analgesic activity was studied in tests by (Chang et al., 2002; Koster et al., 1959; CYTED, 1996; Vogel, 2002; Barzaga et al., 2005; Bansal et al., 2010). Animals treated with the aqueous and hydroalcoholic extracts of *Corynaea crassa,* with doses of 400 mg/kg administered orally and with acetylsalicylic acid, significantly decreased the number of abdominal contortions caused by the analgesic agent acetic acid.

The best results of the analgesic effect were presented for the animals treated with the reference substance (ASA with 90.40% inhibition of contortions abdominal), however, the extracts tested also showed evidence as analgesics under the conditions tested, with inhibition percentages of abdominal contortions greater than 50%. From the statistical point of view, they showed a similar behavior (without significant differences) regarding the decrease in the number of abdominal pains.

Aphrodisiac Activity

The procedure (Yakubu et al., 2007; Tinco et al., 2010; Jung et al., 2017; Ofeimum, 2017; Tajuddin et al., 2004) was followed with some modifications in the observation times and parameters to be measured.

In the experimental animal groups, to which the extracts of the studied plant were administered, it was observed that the sexual behavior of the male rats presented greater frequency in sex drive compared to the negative controlled group. This negative controlled group was only administered a placebo (saline solution). The groups treated with doses of 200 mg/kg with the aqueous extract and sildenafil showed the highest values, with a similar behavior. When the hydroalcoholic extract was administered with a dose of 200 mg/kg, they did not show significant differences in this parameter, but they compared to the previous groups.

Mounting latency was significantly lower in the positive controlled groups and the groups that received the extracts compared to the negative control. The aqueous extracts resulted in behavior similar to the sildenafil group compared to hydroalcoholic extracts.

With another parameter, frequency of intrusions that measures sexual potency (Rampin et al., 2003), all groups showed higher values than the negative control. The extracts tested had a similar behavior without significant differences, and compared to sildenafil, being able to assume that there has been a positive effect on sexual behavior in rats.

Another parameter of sexual behavior evaluated was the latency of interferences where the groups that received the sildenafil and the aqueous extract manifested the shortest time without differences between them. The highest intrusion latency time was for the negative controlled group. A decrease in the latency of intrusions is an indicator of sexual activation, motivation and power (Padashetty and Mishra, 2007). Regarding the ejaculatory frequency, the highest value was achieved for sildenafil followed by the aqueous extract, followed by the hydroalcoholic extract and different from the negative control, which manifested the lowest frequency.

A decrease in ejaculatory latency was observed when the treatments were administered compared to the negative control, being less for the group that received sildenafil followed by the groups that received the aqueous extract, which had a similar behavior. The animals treated with the hydroalcoholic extract presented high values in this parameter without significant differences. In general terms, it can be assumed that the extracts, together with the reference drug, induce an increase in potency and the number of penetrations, reaching ejaculation much more quickly (Padashetty and Mishra, 2007).

Genital screening: In this study, an increase in this parameter was observed in the groups that received sildenafil and extracts compared to the negative control. This test reveals that the riding frequency, intrusion frequency, ejaculatory frequency and genital sniffing have a directly proportional relationship (Granger et al., 1995). The study developed provides evidence of the use of the species *Corynaea crassa* in ethnomedicine as an aphrodisiac. The extracts revealed positive activity on the sexual behavior of the rats tested, where the dose of 200 mg/kg of aqueous extract, making a general consensus of all the variables studied, were those that showed the highest activity, with a behavior comparable to sildenafil. The effect observed in all extracts could be due to the presence of constituents such as saponins, flavonoids, tannins, triterpenoids and steroids that are reported to be present in the plant, and these may have exerted their effect through central and/or peripheral mechanisms.

Nitric oxide values were also determined by the Griess method, expressed as nitrites in plasma and in the corpora cavernosa of the rat penis (Organization for Economic Cooperation and Development (OECD, 2002)). The plasma nitric oxide level in all the samples tested was higher than in the negative control group, being higher for the group treated with the drug sildenafil (7.47 μM), followed by the aqueous extract (6.36 μM), while the group that received the hydroalcoholic extract showed comparable nitric oxide concentrations without significant differences (5.46 μM). The results are according to the different parameters measured in the sexual behavior of the rats where the positive control and the aqueous extracts were the ones that recorded the best responses.

Nitric oxide levels in the corpora cavernosa also increased significantly compared to the negative control in the groups tested. The highest concentration was obtained in the animals treated with sildenafil (23.50 μM), followed by the aqueous extract (21.30 μM), followed by the hydroalcoholic extract (18.61 μM), the same ones that did not show significant differences.

Orally administered *Corynaea crassa* extracts at a dose of 200 mg/kg have been experimentally shown to promote sexual activity in rats under the conditions tested, increasing nitric oxide levels in plasma and in the corpora cavernosa. The ability of the extracts to increase the concentration of nitric oxide in the corpora cavernosa can help explain its aphrodisiac potential. In contrast, sildenafil exerted a more potent action than the extracts, an understandable aspect considering that it is a drug, and the extracts are mixtures of metabolites.

ACUTE ORAL TOXICITY

With the use of the aqueous and hydroalcoholic extracts with the maximum dose of 2,000 mg/kg of weight of the rats, none of the rats died. No clinical signs were reported in any of the animals studied. The Acute Oral Toxicity of the extracts administered to rats orally was determined. It was developed according to the methodology described by the World Health Organization Quality control methods for medicinal plant materials (OECD, 2002; WHO, 1998). Regarding body weight for days 1, 7 and 14 of the experience that corresponds to the dose of 2,000 mg/kg, it could be seen that the treated groups had weight gain during all the weighing carried out, which suggests absence of systemic toxic effects. The samples taken from the selected organs did not show macroscopic affectations, so the professional pathologist decided not to take them for histopathological study in animals.

CONCLUSION

As there is very little background study on these species and none on the native from Ecuador, everything that is reported regarding this species is new, however the following results should be highlighted: I) Through studies, the taxonomy of the species from Ecuador was determined, as well as its macro and micromorphological characteristics; II) The physical-chemical quality parameters of the drug and the aqueous and hydroalcoholic extracts are reported for the first time according to two places of origin; III) New components for the species and genus will be identified using chromatographic and spectrometric methods and IV) Biological activities and non-acute oral toxicity of aqueous and hydroalcoholic extracts are reported for the first time.

REFERENCES

Angulo, A, Rosero, R, Gonzalez, M. 2012. Estudio etnobotánico de las plantas medicinales utilizadas por los habitantes del corregimiento de Genoy, Municipio de Pasto, Colombia. Scielo, Vol. 17, 97–111.

Armijos, C, Cota, J, Gonzalez, S. 2014. Traditional medicine applied by the Saguro yachakkuna: A preliminary approach to the use of sacred and psychoactive plant species in the southern region of Ecuador. *J Ethnobiol Ethnomed.* Vol. 10, 26.

Bansal, P, Gupta, V, Acharya, M, Kaur, H, Bansal R, Sharma, S. 2010. Garlic potential substitute to synthetic aphrodisiacs for erectile dysfunction. *J Pharm Res.* Vol. 2, 37.

Barzaga, F, Nuñez, F, Aguero, F, Chavez, H, Gonzalez, S, Iser, V. 2005. Analgesic effect of the lyophilized aqueous extract of *Ocimum tenuiflorum* L. Cuba. *Cuban Magazine Plant Med.* Vol. 10 (1).

Brako, L, Zarucchi, J L. 1993. Catalogue of the flowering plants and gymnosperms of peru. *Bot Missouri Bot Gard.* Vol. 42, 761–772.

Bussmann, R Glenn. 2010. Medicinal plants used in Northern Peru for reproductive problems and female health. *J Ethnobiol Ethnomed.* Vol. 6, 1–30.

Bussmann, R Sharon. 2006. Traditional medicinal plant use in Northern Peru: tracking two thousand years of healing culture. Peru. *J Ethnobiol Ethnomed.* Vol. 2, 1–18.

Bussmann, R, Glenn, A, Sharon, G, DIaz, D, Pourmand, B, Jonat, S, Somog, G, Guardado, C, Aguirre, R, Chan, K, Meyer, A, Rothrock, A. 2011. Townesmith proving that traditional knowledge works: the antibacterial activity of Northern Peruvian medicinal plants. Peru. *Ethnobot Res Appl.* Vol. 9, 1–30.

Caballero-Serrano, V, McLaren, B, Carrasco, J, Alday, J, Fiallos, L, Amigo, J, Onaindia, M. 2019. Traditional ecological knowledge and medicinal plant diversity in Ecuadorian Amazon home gardens. *Glob Ecol Conserv.* Vol. 17, e00524, ISSN 351–9894, https://doi.org/10.1016/j.gecco.2019.e00524.

Catalogue of the Flowering Plants and Gymnosperms of. Peru *Monogr. 2014. Syst. Bot. Missouri Bot.* Colombia: catalagovirtualparamo@eia.edu.co.

Chang, C, Yang, M, Wen, H, Chem, J. 2002. Estimation of total flavonoid content in propolis by two complementary colorimetric methods. *J Food Drug Anal.* Vol. 10 (3): 178–182.

Chlopicka, J, Pasko, P, Gorinstein, S, Jedryas, A, Zagrodzi, P. 2012. Total phenolic and total flavonoid content, antioxidant activity, and sensory evaluation of pseudocereal bread LWR-Food. *Sci Technol*. Vol. 46, 548–555.

CYTED. Programa Iberoamericano de Ciencia and Tecnología para el desarrollo. 1996. *Research Techniques Manual Ibero-American Program of Science and Technology dor Development*. Sub Programax Pharmaceutical Fine Chemistry, 4–21.

Ditjen POM. Materia medika Indonesia. *Jilid VI*. 1995. *Jakarta*. Republica Indonesia: Departement Kesehatan, 103–113.

Fernandez, R, Cruzado, L, Bonilla, R, Ramirez, C, Toche, T, Curay, C. 2017. Identification of secondary metabolites and anti-inflammatory effect of the ethanolic extract of *Chromolaena leptocephala* (DC) R.M. King & H. Rob "Chilca negra". *Peruvian J Integrat Med*. Vol. 2 (3), 779–784.

Gattuso, M, Gattuso, S. 1999. *Procedures manual for drug power analysis*. Urquiza Argentina: Editorial of the National University of Rosario. ISBN 950-673-199-3.

Gómez, L. 1983. Family 61 Balanophoraceae. In W. Burger. *Costaricensis Fieldiana Bot*. Vol. 13, 89–93.

Gonzalez, G, Ospina, G, Rincon, V. 2011. Anti-inflammatory activity od extracts and fractions of *Myrcianthes leucoxila, Calea prunifolia, Curatella americana* and Physalis peruvian in models of atrial edema due to ATP, plantar edema due to carrageenan, and collagen-induced arthritis. *Biosalud*. Vol. 10 (1), 9–18.

Granger, Donald L, Taintor, Read R, Boockvar, Kenneth S, Jr., Hibbs, John B. 1995. Determination of nitrate and nitrite in biological samples using bacterial nitrate reductase coupled with the Griess reaction. *Methods*. Vol. 7 (1), 78–83, ISSN 1046-2023, https://doi.org/10.1006/meth.1995.1011.

Guo, J, Zhang, D, Yu, C, Yao, L, Chen, Z, Tao, Y, Cao, Wa. 2019.Phytochemical analysis, antioxidant and analgesic activities of *Incarvillea compacta* Maxim from the Tibetan Plateau. *Molecules*. Vol. 24, 1692.

Hansen, B. 1983. Balanophoraceae in G W Harling & B B Sparre Fl. Ecuador. Goterborg & Stockholm: University of Goteborg & Swedish Museum of Natural History, Vol. 23, 1–80.

Jung, M, Oh, K, Choi, E, Kim, Y, Bae, D, Oh, D. 2017. Effect of aqueous extract of D*endropanax morbifera* leaf on sexual behavior in male rats. *J Food Nutr Res*. Vol. 5, 518–521.

Knapp, S. 2012. Missouri Botanical Garden obtained of *Corynaea crassa* Hook. f. htpp://www.tropicos.org/name/03000026?projectid=3

Koster, R, Anderson, M, De Beer, E. 1959. Acetic acid for analgesic screening. *Fed Proc*. Vol. 18, 412–417.

Kuntze Hansen, B. 1993. *In G. W. Harling, B. B. Spare Fl. Ecuador*. Goteborg & Stockholm: University of Goteborg & Swedish Museum of Natural History.

Lobo, S. 2003. The hosts of the hemipparasitic plants of the Loranthacear family in Costa Rica. *National Herbarium of Costa Rica Jankesteriana*. Vol. 6, 17–20.

Lopez, A, Gutierrez, Y, Miranda, M, Choez, I, Ruiz, Scull, R. 2021. Pharmacognostic, phytochemical and anti-inflammatory effects of *Corynaea crassa:* A comparative study of plants from Ecuador and Peru. *Pharmacogn Res*. Vol. 12, 394–402.

Malca, G, Henning, L, Sieler, J, Bussmann, R. 2015. Constituents of C*orynaea crassa* "Peruvian Viagra". *Brazilian J Pharmacogn*. 25 (2), 92–97.

Miranda, M, Cuellar, A. 2000. *Laboratory practice manual pharmacognosy and natural products*. Havana: Poligrafica Felix Varela, Vol. 1, 48–50.

Norma Ramal NRSP 309. 1992. *Norma Ramal Raw drug. Test methods*. Medicines of plant origin, 16–20.

Ofeimum, J, Avinde, B. 2017. Preliminary investigation of the aphrodisiac potential of the methanol extract and fractions of Rhaphiostylis beninensis Planch ex Benth (lcacinaceae) root on male rats. *J Sci Pract Pharm*. 4 (1): 182–188.

Organization for Economic Cooperation and Development (OECD). 2002. *Guidelines for testing of chemical*. Paris: OECD Publishing, Vol. 201, 1–21.

Pachacuti-Yamqui, J De Santa C. 1992. *One of the plants uses in the aphrodisiac mixtures is commonly known as chutarpo or huanarpo as Corynaea crassa Hook f. List of antiquities of this Kingdom of Peru various*. Antiquities of Peru. Peru. Chronicles of America, Vol. 70, 16.

Padashetty, S, Mishra, R, Boockvar, K, Hibbs, J. 1996. Determination of nitrate and nitrite in biological samples using bacterial nitrate reductase coupled with the Griess reaction. *Methods Companion Meth Enzymol.*

Padashetty, S, Mishra, S. 2007. Aphrodisiac studies of *Tricholepis glaberrima* with supportive action from anti-oxidant enzymes. *Pharmaceut Biol.* Vol. 45, 580–586.

Peacock, H, Bradbury, S. 1973. *Peacock¨s elementary microtechnique*. London: Edward Arnold, 4a edic.

Pourmorad, F, Hosseinimehr, S, Shahabimaid, N. 2006. Antioxidant activity, phenol and flavonoid contents of some selected Iranian medicinal plants. *Afr J Biotechnol*. Vol. 5 (11), 1142–1145.

Rampin, O, Jerome, S, Suaudeau, C. 2003. Proerectile effect of apomorphine in mice. *Life Sci*. Vol. 72 (21), 2329–2336.

Sato, H, Gonzalez, M. 2013. Floral development and anatomy of pistillate flowers of Lophophytum (Balanophoracea), with special reference to the embryo sac inversion. *Flora*. Vol. 219, 35–47.

Tajuddin, A, Latif, A, Qasmi, I. 2004. Effect of 50% ethanolic extract of *Syzygium aromaticum* (L) Merr & Perry (Clove) on sexual behavoiur of 72 normal male rats. *BMC Complement Altern Med*. Vol. 4, 17.

Tene, V, Malagon, O, Vita, P, Vidari, G, Armijos, Ch, Zaragoza, T. 2007. An ethnobotanical survey of medicinal plants used in Loja and Zamora-Chinchipe, Ecuador. *Journal of Ethnopharmacology*, Vol. 111 (1), 63–81.

Tinco, J, Arroyo, J, Bonilla, P. 2010. *Modulating effect of erection by methanolic extract of Jatropha macrantha Mull. Arg "Huanarpo macho" in rats with induction of erectile dysfunction*. Mayor University of San Marcos,. Vol. 72 (3), 161–168.

Tropicos. 2007. Nomenclatura Database. *Missouri Botanical Garden*. http://www.tropicos.org/

Tropicos. 2019. Name Corynaea crassa Hook. f. www.tropicos.org/name/03000026?projectid=

Tupac, O, Mora, M, Costa J. 2009. First host record for the root parasite *Corynaea crassa* (Balanophoracea). *Scielo Bogota*. Vol. 14 (3), 199–204.

Udupa, S, Udupa, A, Kulkami, D. 1994. Anti-inflammatory and wound healing properties of *Aloe vera*. *Fitoterapia*. Vol. 65 (2), 141–145.

Vogel, H, Vogel, W. 2002. *Writing tests. Drug discovery and evaluations Pharmacological assays*. New York: Springer-Verlag Berlin.

WHO. 1998. *World Health Organization Quality control methods for medicinal plant materials*. Ginebra: WHO/PHARM/92.559.

Winter, C, Risley, E, Nuss, G. 1962. Carrageenin-induced edema in hind paw of the rat as an assay for anti-inflammatory drugs. *Proc Soc Exp Biol Med NYN*. Vol. 111, 544–547.

Yakubu, M, Akanji, M, Oladiji, A. 2007. Male sexual dysfunction and methods used in assessing medicinal plants aphrodisiac potentials. *Pharmacognosy Rev*. Vol. 1 (1), 49–56.

Zambrano, L, Buenaño, M, Mancera, N, Jimenez, E. 2015. Estudio etnobotánico de plantas medicinales utilizadas por los habitantes del área rural de la Parroquia San Carlos, Quevedo, Ecuador. Scielo. Vol. 17, 1.

10 Smilax Species Used in Traditional Medicine of Ecuador. Chemical Composition and Pharmacological Properties

Pilar Soledispa, Raisa Mangas, Glenda Sarmiento and Migdalia Miranda Martínez

CONTENTS

DOI: 10.1201/9781003173991-13

INTRODUCTION

Smilax (Smilacaceae), a very diverse genus, comprises about 250 species of flowering plants, many of which are woody or thorny. It is located in temperate, tropical and subtropical zones of the world (Huft 1994; Judd et al. 2002). A wide variety of medicinal properties have been attributed to this genus: sudorific, diuretic, purifying, laxative and hypoglycemic, among others (Ocampo 1994, Gupta 1995). The most used parts are the rhizome, the leaves and the stem, which are usually used as an infusion. They have also been used by the food industry as a flavoring (MacVean 2016). Its chemical composition is varied, being identified as steroidal saponins, phytosterols, triterpenoids, flavonoids, alkaloids, tannins and coumarins in different plant organs (Cáceres et al. 2011; Álvarez et al. 2016; González et al. 2017). Currently, products made from this genus (*Smilax aspera* and *Smilax ornata*) are marketed from the processing of the root and rhizome, in presentations as beverages with anti-inflammatory, antirheumatic and diuretic properties, and to treat psoriasis (Méndez 2018).

The species *Smilax purhampuy* R., native to the Amazon, grows in pantropical-subtropical zones, being widely distributed throughout the world, mainly in America, Europe and East Asia (Téllez 1996). In our area it is located in Peru, Nicaragua, Bolivia, Colombia, Ecuador, Costa Rica, Venezuela, Brazil and Honduras (Rivas et al. 2017). Its common names in Latin America include cuculmeca, stick of life, dog grape, sarsaparilla, sarsaparilla, zarza (JR Global del Perú SAC 2011; Jiménez 2014). It is known for some of its healing and therapeutic properties such as the elimination of excess cholesterol, triglycerides, arthritis, intestinal, stomach and prostate inflammation, chronic gastritis and cysts (JR Global del Perú SAC 2011). Investigations of this species are scarce, comprising only the study of the leaves where the physical chemical parameters of the drug were established, as well as the presence of metabolites suggested by phytochemical screening; others with possible pharmacological and anti-inflammatory effects were identified (Parrales and Villamar 2018).

DEFINITION

Smilax purhampuy Ruiz (Figure 10.1), genus *Smilax* L., is a species from the family Smilacaceae Vent found in Ecuador. They are hairless climbing plants belonging to the order Liliales (Bisby et al. 2010; Ferrufino 2014).

SYNONYMS

S. panamensis is the same taxon known as *S. febrifuga* Kunth. Since 2012 the World Checklist of Selected Plant Families reports that *Smilax Purhampuy* Ruiz is an accepted name (record 288875) as a replacement for *S. febrifuga* (Bisby et al. 2010; WFO 2021).

This species is known as *Smilax febrifuga* var. aequatoris A. DC. in Ecuador, *Smilax poeppigii* Kunth or *Smilax insignis* Kunth in Peru, *Smilax panamensis* Morong in Panamá, *Smilax ramonensis* F. W. in Costa Rica, *Smilax graciliflora* A. C. Sm. in Brazil (Ferrufino 2014; WFO 2021).

FIGURE 10.1 The plant *Smilax purhampuy* Ruiz collected in Ecuador. (Photo by author)

SELECTED VERNACULAR NAMES

Zarzaparrilla, Sarsaparilla, Zarza, Zarza Masha, Chinese liana, Chinese root or Chinese vein (Ferrufino 2014).

GEOGRAPHICAL DISTRIBUTION

Figure 10.2 illustrates the *Smilax* locations where they have been reported, being the Pastaza, Napo and Orellana Provinces located in the Amazon Region, the most representative locations including part of the slopes of the Andes, to the Amazonian plains of Ecuador. These represent geographic regions such as evergreen seasonal forest, open areas, riparian forests, montane rain forests, 0–800 m.

Additionally, these species of *Smilax* can be found in Peru, Venezuela, Colombia, Brazil, French Guiana, Panama and Costa Rica (Ferrufino 2014).

DESCRIPTION

It is a perennial climbing plant with a very stout branched underground root that produces stems from the same root, rounded with straight proximal yellowish-green spines. The membranous leaves of 5–7 submarginal veins have ribbing connected by reticulate veins, elliptical to elliptical lanceolate, 15–20 × 8–12 cm, acute base, acuminate apex, entire margin. It has long, spiraling opposing tendrils. Flowers and fruits are not seen at harvest (Mero and Muñoz 2019).

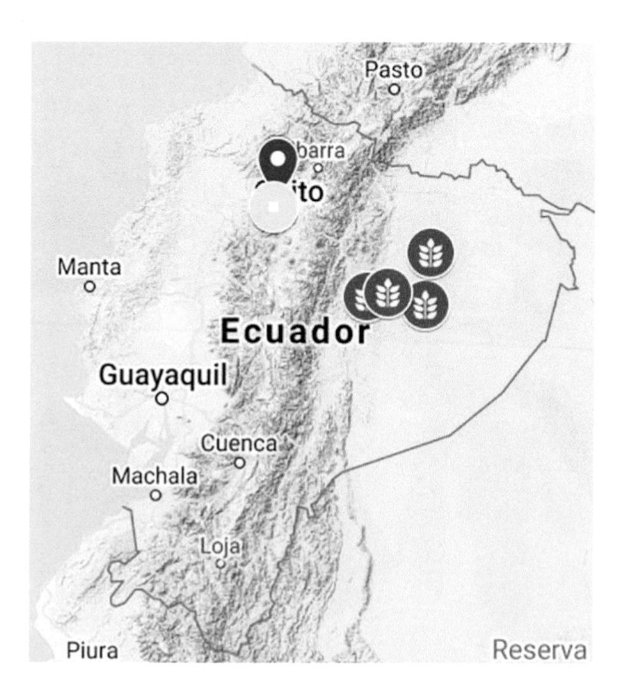

FIGURE 10.2 Distribution of *S. purhampuy* (red), *S. fluminensis* (yellow) and *S. spinosa* (blue). (Prepared by the author using Google My Maps)

PLANT MATERIAL OF INTEREST: DRIED RHIZOME

General Appearance

The rhizome is short tuberous, whitish-brown on the outside and yellowish-white on the inside, with a sprained and branched shape with a size of 9–20 cm long and a diameter of 3–7.5 cm (Mero and Muñoz 2019).

Organoleptic Properties

Odor, sui generis; taste, very bitter; color, yellowish-brown to yellowish-white (Grieve 2013; De la A and Pilligua 2018).

Microscopic Characteristics: Powdered Plant Material

The micromorphological analysis of the powdered drug of the rhizomes of the species is shown in (Figure 10.3), which allowed to observe (A) a set of parenchyma cells of variable size, and (B) xylem vessels with numerous sliced thickenings, frequently associated with other elements of the conductive tissue. Elongated, spindle-shaped and pointed structures (C) were also seen, corresponding to the so-called fibers, which could suggest a type of filiform sclereid. In another sample of the powdered drug (D), xylem vessels with rimmed pits were seen. When performing the histochemistry with Lugol's reagent (E), starch granules were verified as a remarkable shape, which took on a blackish coloring.

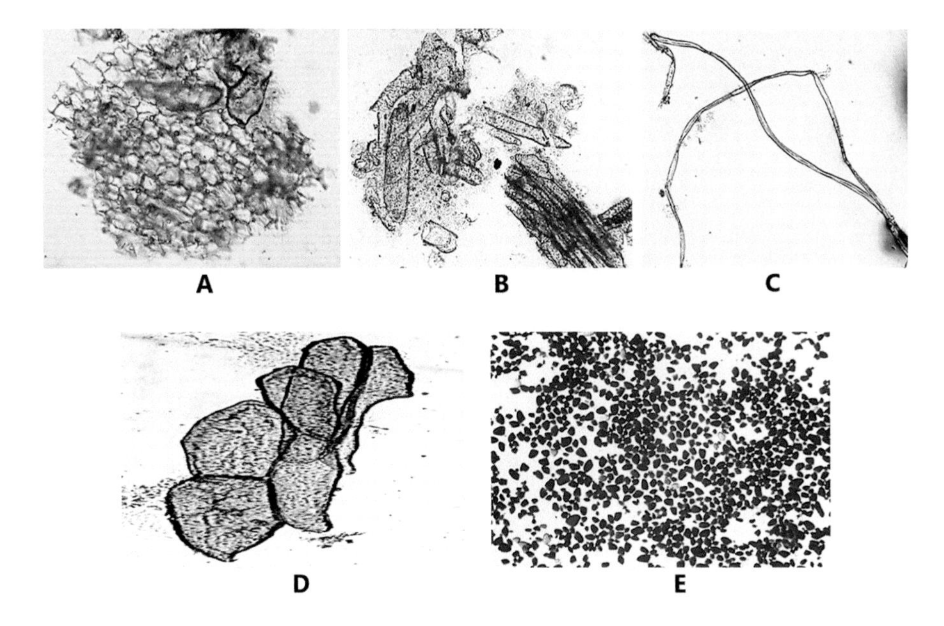

FIGURE 10.3 Micromorphology of *Smilax purhampuy* rhizomes powder drug.

GENERAL IDENTITY TESTS

Regarding macroscopic, microscopic and chemical composition, thin-layer chromatographic analysis indicated the presence of phenolic and terpenic structures (Khandelwal 2008; Miranda and Cuéllar 2000).

PURITY TESTS

Total Ash

It was 1.32%, which is below the admitted limit of 5.0% (Lou 1980; WHO 1998); the Chinese Pharmacopoeia refers up to 15% (Commission 2015).

Humidity Residues

The moisture content was 8.16%, which was framed within the range established by the literature of 8 to 14% for medicinal plants (WHO 1998; Miranda and Cuéllar 2001).

Alcohol-Soluble Extractive

The strongest chemical extraction was achieved with the 50 and 80% ethanolic extract.

Quality Physical-Chemical Parameters

The quality determinations were carried out on the 80% hydroalcoholic extract by means of the procedure described (Miranda and Cuéllar 2000; NRSP-312 1992), 3 replicates were made for each experiment.Table 10.1 shows parameters being evaluated:

TABLE 10.1
Physicochemical Parameters of Rhizome Extract

Parameters	Results $\bar{x}$/SD
pH	6.27/0.02
Total solids (%)	0.89/0.04
Refractive index (g/mL)	1.3580/0.0003
Relative density (g/mL)	0.8743/0.0011
Capillary analysis (height in cm)	8.3/0.14

$\bar{x}$/*SD:* mean/standard deviation (n=3)

The Organoleptic properties presented a reddish coloration and characteristic odor. The refractive index was performed with an ABBE digital refractometer and the relative density by pycnometry.

Other Purity Tests

The ashes insoluble in acid and soluble in water were less than 1%, which are within the ranges required by the WHO of less than 2% for medicinal plants (WHO 1998, 34; Miranda and Cuéllar 2001).

CHEMICAL ASSAYS

The phytochemical screening of the hydroalcoholic extract contains lactones/coumarins, triterpenoids/steroids, tannins/phenols, quinones, flavonoids, anthocyanidins, reducing substances and alkaloids.

In the thin layer chromatography (TLC), it was observed that some spots had a characteristic behavior of phenolic and terpenic structures, all of them assayed by means of spectrophotometric techniques (Khoddami et al. 2013), determining a total phenols content of 2.73 mg/mL and total flavonoids of 0.55 mg/mL.

MAJOR CHEMICAL CONSTITUENTS

The rhizome contains dihydrocorinantein (Figure 10.4), which was the second major compound but with a better resolution and a relative abundance of 2.83% related to indole alkaloids. Other major constituents were the monosaccharide L-altrose (2.47%) and the major component stearic acid (3.97%), all of them assayed by means of gas chromatography coupled with mass spectrometry.

Other constituents are hexadecanoic acid (palmitic acid) (1.65±0.01%), followed by butanedioic acid (succinic acid) (1.163±0.015%), citric acid (0.32%) and inositol (0.31±0.02%), among others.

DOSAGE FORMS

The pharmaceutical forms used are tablets, syrups and, most commonly, extracts by means of powder or crude plant material. Their extraction methods are by maceration

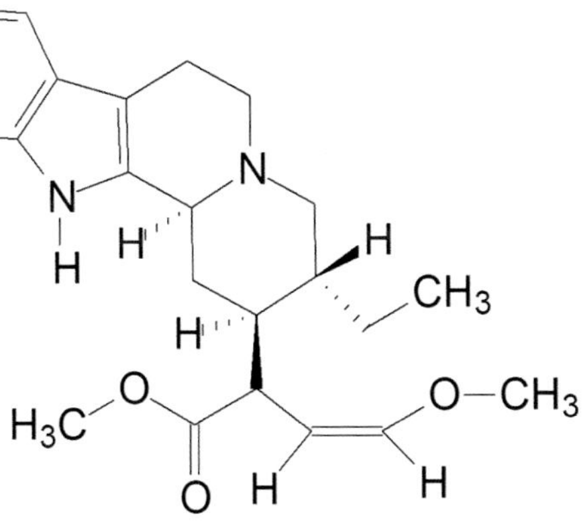

FIGURE 10.4 Dihidrocorinantein, chemical compound of rhizome *Smilax purhampuy*.

and decoction. They must be stored in a well-ventilated dry environment protected from light (ARCSA 2018).

MEDICINAL USES

Uses Supported by Clinical Data

None reported scientifically.

Uses Described in Pharmacopoeias and in Traditional Systems of Medicine

To manage syphilis and other skin conditions (Wilson 2016; Howes 2018), for Urticaria (Bodemer 2018) and Psoriasis (Hou et al. 2011). The rhizome is also used as herbal treatment of Urinary Tract Infection (Loo 2009) and Rheumatoid Arthritis (Dinesh and Rasool 2019; Berman et al. 2015).

Uses Described in Folk Medicine, Not Supported by Experimental or Clinical Data

The extracts of rhizome in pieces are commonly used in the Amazon of Ecuador as a treatment for skin problems, kidney conditions, intestinal disorders, polycystic ovary, depurative, arthritis and venereal diseases (Mero and Muñoz 2019; De la A and Pilligua 2018).

PHARMACOLOGY

Experimental Pharmacology Anti-Inflammatory Activity

Research recommends the use of medicinal plants as promising results have been shown in the prevention and treatment of various pathologies, including inflammation. Besides, they have low cost and minimal side effects (Ginwala et al. 2019; Zhang et al. 2019).

The hydroalcoholic extracts of rhizomes (80% ethanol) and leaves (50% ethanol) of *Smilax purhampuy* (intragastric administration: 400 mg/kg) demonstrated their anti-inflammatory efficacy by means of the experimental model of plantar edema by carrageenan-induced (subcutaneous: 0.3 mL of carrageenan 3%) in Wistar albino rats (females: 180–210 g, 4 groups of 6 animals each).

Its administration caused a significant reduction ($P < 0.05$) of the paw edema; it was observed (Figure 10.5) how the leaf extract had a comparable behavior with indomethacin (10 mg/kg) used as a positive control, with inhibition percentages by up to 53%, followed closely in a lower percentage by the rhizome extract in 5 hours, showing in both cases the anti-inflammatory power under the conditions tested.

The demonstrated pharmacological effect could be related to phenolic compounds, especially flavonoids, detected in phytochemical screening and TLC, which inhibit the inflammatory response and comprise the main class of compounds present in the *Smilax* genus.

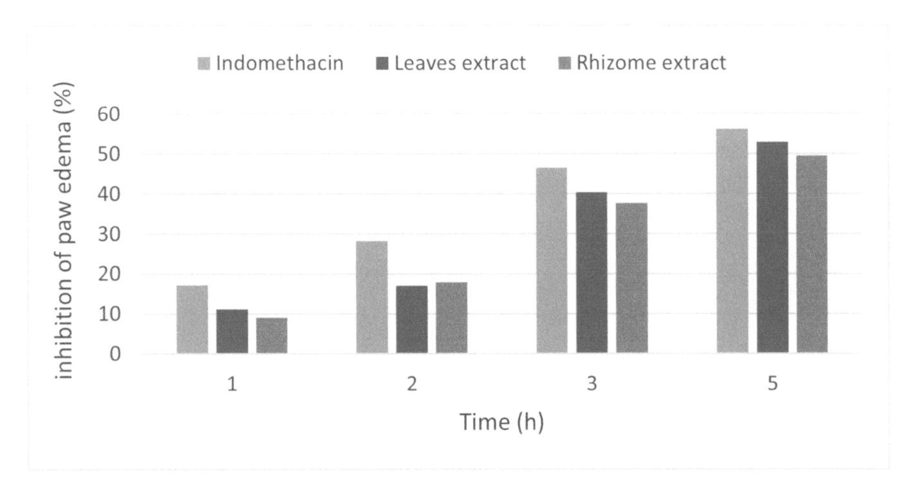

FIGURE 10.5 Effects of *S. purhampuy* extracts on carrageenan-induced plantar edema in rats.

Antioxidant Activity

Antioxidants are classified according to their origin – in this case, exogenous origin. Among them is vitamin C or ascorbic acid, α-tocopherol or vitamin E, beta-carotenes and polyphenols (Sen et al. 2010). Polyphenols, including simple phenols, benzoic and cinnamic acids, coumarins, tannins, lignans and flavonoids account for the antioxidant richness of medicinal plants (Khoddami et al. 2013; Silva et al. 2018).

The antioxidant activity was evaluated in the ethanolic extract (80%) of the rhizomes by means of the experimental methods *in vitro*: scavenging capacity of radicals 2,2'-azinobis (3-ethylbenzothiazoline) -6-sulfonic acid (ABTS) (100, 200, 300, 400 and 500 µg/mL) and 2,2-diphenyl-1-picryl hydracil (DPPH), as well as the ferric-reducing antioxidant power assay (FRAP), the last two experiments in concentrations of 25, 50, 100, 150 and 200 µg/mL. Figures 10.6–10.8 illustrate the results of these experiments.

The results of the 3 methods verified that the extract had a dose-dependent behavior, manifesting a high antioxidant activity due to the inhibition capacity of the DPPH and ABTS radicals, as well as of its ferro-reducing activity with high values of µM equivalents of ascorbic acid and $FeSO_4$ as the concentration increased. The extract at the maximum concentration tested by the DPPH method showed a similar behavior to the two reference substances, with inhibition percentages greater than 84%. The extract at the minimum concentration tested by the ABTS method had a sequestration capacity greater than 50% than vitamin C and Trolox, used as positive controls, while, at concentrations of 200, 300 and 500 µg/mL, the extract showed a higher percentage of inhibition than Trolox. This considerable antioxidant effect could be attributed to an appreciable concentration of phenolic compounds.

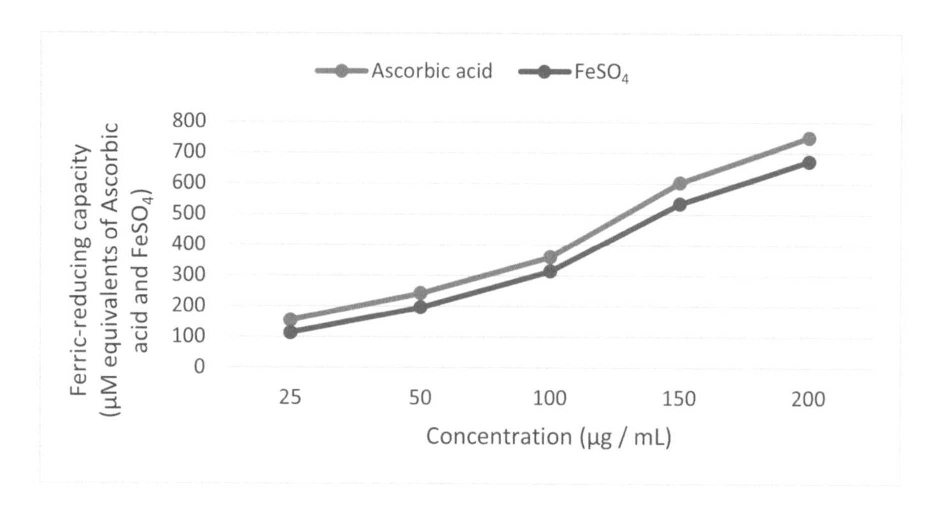

FIGURE 10.6 Ferric-reducing capacity of the ethanolic extract of rhizomes.

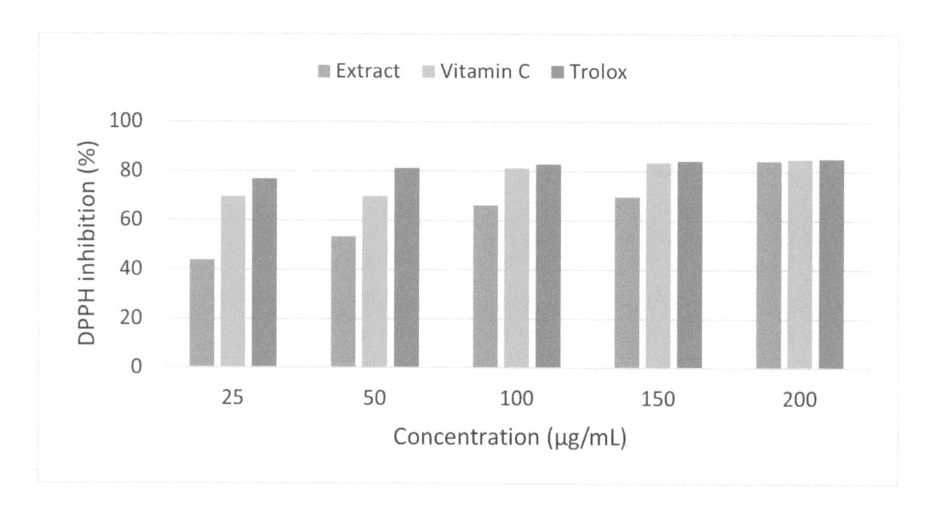

FIGURE 10.7 Scavenging capacity of DPPH radical of the ethanolic extract of rhizomes and the reference substances.

Antimicrobial Activity

The ethanolic and aqueous extracts of the rhizome of the species were evaluated (50 µl of extract in dimethylsulfoxide at 100, 75, 50 and 25%), by means of *in vitro* experimental methods: Mueller Hinton Agar, Kird-Bauer and Duraffourd scale. Two gram-positive bacteria, *Staphylococcus aureus* (ATCC 25923) and *Bacillus cereus* (ATCC 7464), and the fungus *Candida albicans* (ATCC 10231) were used as microorganisms and for the positive control; vancomycin, vistatin and sulfamethoxazole were used.

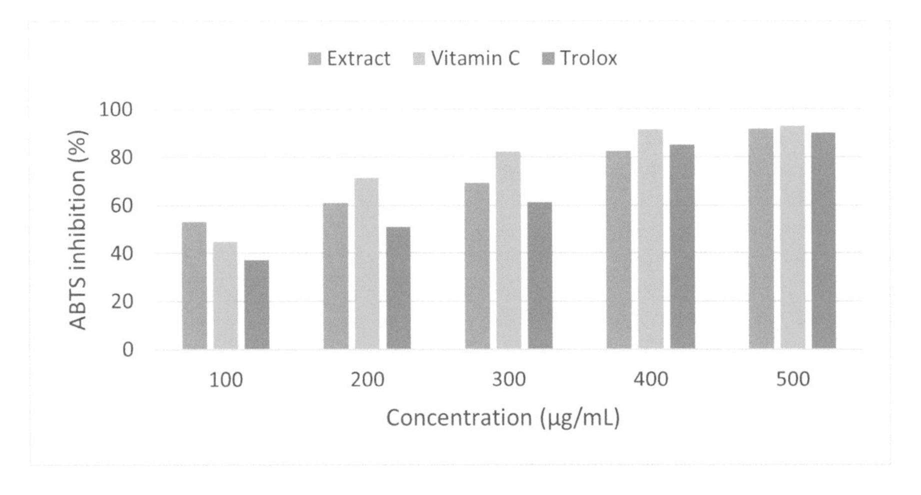

FIGURE 10.8 Scavenging capacity of ABTS radical of the ethanolic extract of rhizomes and the reference substances.

The averages of the halo inhibition of the microorganisms were related to the antibiotics. In the case of *Staphylococcus aureus*, the measurement of sulfamethoxazole was 31 mm, for *Bacillus cereus* the measurement of vancomycin was 21 mm and lastly for the yeast *Candida albicans* the measurement of antifungal nystatin resulted in 25 mm. In the ethanolic extract the zone of inhibition varied from 10–16 mm for *S. aureus* and *B. cereus* and 7–10 mm for *C. albicans* within 48 hours. According to the Duraffourd scale, the extracts would be sensitive from the concentration of 25% and 50% and very sensitive at 75% and 100%.

It was demonstrated that the ethanolic extract has an inhibitory power very similar to vancomycin, observing a greater halo inhibition for the *Staphylococcus aureus* strain, followed by the *Bacillus cereus* and *Candida albicans* strain, unlike the aqueous extract that did not present antimicrobial activity in any of them. The inhibitory effect suggests being related to phenols and flavonoids found in the rhizome.

Normoglycemian Activity

The administration of the ethanolic extract of the rhizomes (oral route: 250 and 1,000 mg for 7 days) through the induction of type II diabetes mellitus by alloxane (intraperitoneal route: 0.5 mL, 100 mg/kg) in albino male mice, (age: 3–4 weeks; average weight: 25 g; 5 groups of 4 animals according to weight) caused a significant decrease in plasma glycemic values (Figure 10.9), very similar to the control group (glibenclamide 5mg). The concentration of 250mg (group E) achieved a greater reduction than that of 1,000 mg (group D).

The histopathological study showed exocrine cells in group E with a cytoplasmic turbidity process as well as in the epithelium of the secretory tubule, similar to group C that furthermore did not show alteration in the pancreatic β cells of the islets of

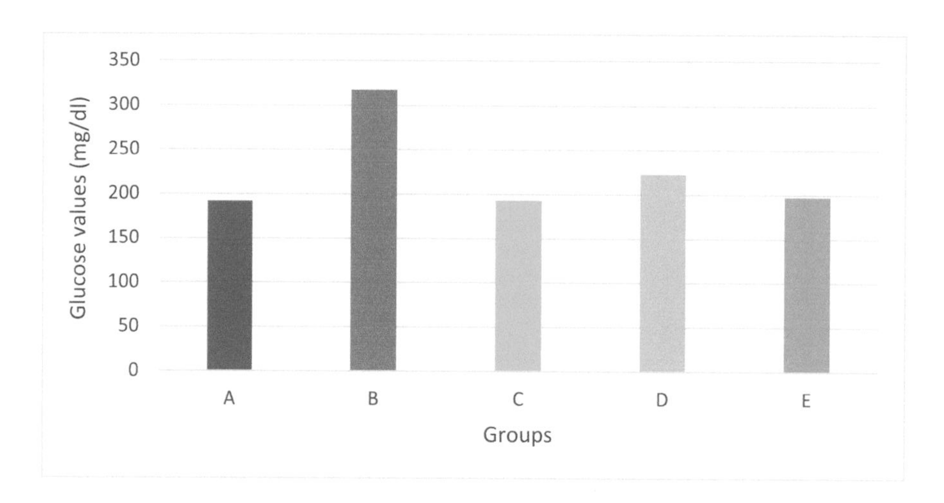

FIGURE 10.9 Normoglycemic effect of the ethanolic extract of the rhizomes. (A) No treatment, (B) Alloxane (0.5ml), (C) Glibenclamide (5mg), (D) Ethanolic extract (1,000mg), (E) Ethanolic extract (250mg).

Langerhans; on the contrary, groups B and D presented alteration in their pancreatic β cells. These results indicate protection against diabetic disease.

It is confirmed that the extract possesses normoglycemic activity and can lower blood glucose and glycosylated hemoglobin (HbA1c) levels, suggesting that isolated phenolic compounds are potentially involved in this effect.

Clinical Pharmacology

None reported scientifically for this species.

CONTRAINDICATIONS

The safety of extracts of the rhizome in pregnancy or lactation has not been established. Therefore, until such data are available the use during pregnancy is contraindicated.

WARNINGS

Because saponins can cause hemolytic symptoms, their prescription should be avoided in anemic patients. No more information available.

PRECAUTIONS

Carcinogenesis, Mutagenesis, Impairment of Fertility

The safety of extracts of the rhizome has not been established with respect to fertility.

Pregnancy: Non-Teratogenic Effects

The safety of extracts of the rhizome has not been established with respect to pregnancy.

Nursing Mothers

Excretion of the rhizome into breast milk, and its effects on the newborn have not been established; therefore, use of the herb during lactation is not recommended.

Pediatrics Use

The safety and efficacy of the rhizome in children have not been established.

Other Precautions

No information is available concerning general precautions, drug and laboratory test interactions, or teratogenic effects on pregnancy.

ADVERSE REACTIONS

No systematic studies have assessed organ function during acute or chronic administration of the rhizome.

POSOLOGY

Maximum daily oral dosage of crude plant material: 1.5–6g.

CONCLUSIONS

The pharmacognostic study of the *Smilax purhampuy* rhizome allowed to establish the main physicochemical quality parameters of the crude drug and the 80% hydroalcoholic extract. The analytical methods to analyze the chemical profile of the extract suggested the presence of several phytoconstituents, including phenolic compounds and terpenes. The GC/MS analysis showed dihydrocorinantein as one of the main constituents, in addition to carboxylic acids. The hydroalcoholic extracts of rhizomes and leaves demonstrated their anti-inflammatory efficacy under the conditions tested. On the other hand, the hydroalcoholic extract of rizhomes showed antioxidant activity being dose-dependent, an inhibitory power very similar to vancomycin, and normoglycemic activity. These results could be attributed to the presence of phenolic compounds. The extract of the rhizome is commonly used to treat skin problems, kidney conditions and intestinal disorders. The safety of the rhizome has not been established.

REFERENCES

Álvarez, S., Hermosilla, E.S., Piña, G.R. 2016. Tamizaje Fitoquímico de extractos en diclorometano, etanol y agua de Smilax havanensis Jacq. (Raíz de China). Revista Granma Ciencia. 20: 1–10. ISSN 1027–975X.

ARCSA. 2018. "Base de datos de productos naturales". Accessed March 16, 2021. www.controlsanitario.gob.ec/wp-content/uploads/downloads/2018/03/Base-de-datos-de-Productos-Naturales-VUE-27-02-2018.xlsx

Berman, B., Lewith, G., Manheimer, E., et al. 2015. Chapter 48A—Complementary and Alternative Medicine, Rheumatology, ed. Mosby. Missouri: Elsevier. https://doi.org/10.1016/B978-0-323-09138-1.00048-6

Bisby, F.A., Roskov, Y.R., Orrell, T.M., et al. 2010. Species 2000 & ITIS catalogue of Life: 2010 annual checklist taxonomic classification. DVD; Species 2000: Reading, Reino Unido. www.catalogueoflife.org/annual-checklist/2010/details/species/id/7277246 (accessed February 2, 2021).

Bodemer, A. 2018. Chapter 74 — Urticaria. In Integrative Medicine, ed. D. Rakel, 739–746. e2. Elsevier. https://doi.org/10.1016/B978-0-323-35868-2.00074-8

Cáceres, A., Cruz, S.M., Martínez V., Gaitán I., Santizo A., Gattuso S., Gattuso, M. 2011. Estudios etnobotánicos, farmacognósticos, farmacológicos y fitoquímicos de Smilax domingensis en Commission Chinese Pharmacopoeia. 2015. Pharmacopoeia of the People's Republic of China. Peking: Chinese Medical Science and Technology Press.

De la A., A.A., Pilligua, J.C. 2018. "Estudio farmacognóstico y fitoquímico, preliminar del rizoma de la Smilax china". BSc. diss., Universidad de Guayaquil.

Dinesh, P., Rasool, M. 2019. Chapter 22 — Herbal Formulations and Their Bioactive Components as Dietary Supplements for Treating Rheumatoid Arthritis. In Bioactive Food as Dietary Interventions for Arthritis and Related Inflammatory Diseases, ed. Academic Press, 385–399. Cambridge: Elsevier. https://doi.org/10.1016/B978-0-12-813820-5.00022-2

Ferrufino-Acosta, L. 2014. Taxonomic revision of the genus Smilax (Smilacaceae) in Central America and the Caribbean Islands. Willdenowia 40 no. 2: 227–280. https://doi.org/10.3372/wi.40.40208.

Ginwala, R., Bhavsar, R., Chigbu DeGaulle, I., et al. 2019. Potential role of flavonoids in treating chronic inflammatory diseases with a special focus on the anti-inflammatory activity of apigenin. Antioxidants 8, no. 35: 30. https://doi.org/10.3390/antiox8020035

González, J.Y., Monan, M., Cuéllar, A., De Armas, T., Gómez, E., Dopico, E. 2017. Pharmacognostic and Phytochemical Studies of Smilax domingensis Willd. in Cuba. Am J Plant Sci. 8: 1462–1470. https://doi.org/10.4236/ajps.2017.86100

Grieve, Margaret. 2013. A Modern Herbal: Volume 1. New York, USA: Courier Corporation. https://books.google.com.ec/books/about/A_Modern_Herbal.html?id=wwzCAgAAQBAJ&redir_esc=y

Gupta; M.P. 1995. Plantas medicinales iberoamericanas. CYTED-SECAB. Ed. Colombia: Presencia Ltda.

Hou, W., Xu, G., Wang, H. 2011. CHAPTER 12 — Psoriasis. In Treating Autoimmune Disease with Chinese Medicine, ed. Churchill Livingstone, 205–216. London: Elsevier. https://doi.org/10.1016/B978-0-443-06974-1.00012-9

Howes, M.-J.R. 2018. Chapter 28 — Phytochemicals as Anti-inflammatory Nutraceuticals and Phytopharmaceuticals. In Immunity and Inflammation in Health and Disease, ed. Academic Press, 363–388. Cambridge: Elsevier. https://doi.org/10.1016/B978-0-12-805417-8.00028-7

Huft, M.J. 1994. Smilacaceae. En: Davidse, G.; Sousa, M:; Arthur, S.; Chater, O. Ed. Flora Mesoamericana. 6–25. The Natural History Museum, Instituto de Biología (UNAM). London: Missouri Botanical Garden Press.

Jiménez, A.E. 2014. Determinación de componentes y capacidad antioxidante mediante GC/MS del extracto de zarzaparrilla (Smilax domingensis Willd) y elaboración de bebida de zarzaparrilla nutracéutica. BSc. diss., Universidad de Guayaquil, Facultad de Ingeniería Química.

JR Global del Perú S.A.C. 2011. *INKAPLUS. Ficha Técnica-Zarzaparrilla* [Online]. Available from: www.inkaplus.com/media/web/pdf/Zarzaparrilla.pdf

Judd, W.S., Campbell, C.S., Kellogg, E.A., Stevens, P.F. 2002. Plants systematics. A philogenic approach. Sinauer: Publisher Sunderland.

Khandelwal, K.R. 2008. Practical Pharmacognosy: Techniques and Experiments. Pune: Pragati Books Pvt. Ltd. 9, 15. https://books.google.com.ec/books/about/Practical_Pharmacognosy.html?id=SgYUFD_lkK4C&redir_esc=y

Khoddami, A., Wilkes, M.A., Roberts, T.H. 2013. Techniques for Analysis of Plant Phenolic Compounds. Molecules 18, no. 2: 2328–2375. https://doi.org/10.3390/molecules18022328

Loo, M. 2009. CHAPTER 61 — Urinary Tract Infection. In Integrative Medicine for Children, W.B. Saunders, 456–462. Philadelphia: Elsevier. https://doi.org/10.1016/B978-141602299-2.10061-1

Lou, Z.-C. 1980. General Control Methods for Vegetable Drugs. Comparative Study of Methods Included in Thirteen Pharmacopoeias: An Proposals on Their Internacional Unification. Ginebra: WHO/PHARM/80.502

MacVean, A.L. 2016. Diversidad, distribución e importancia Económica de Smilax (Smilacaceae) de Guatemala, 1: 163–173.

Méndez 2018. Zarzaparrilla: propiedades, beneficios para la salud. www.nutrioptima.com/guia/hierbas/zarzaparrilla-propiedadesbeneficios/

Mero, A.C., Muñoz, K.X. 2019. "Estudio farmacognóstico y fitoquímico preliminar del rizoma de Smilax purhampuy R." BSc. diss., Universidad de Guayaquil.

Miranda, M. Cuéllar, A. 2000. Manual de prácticas de laboratorio. Farmacognosia y productos naturales. 25–49, 74–79. Ciudad Habana: Félix Varela.

Miranda, M., Cuéllar, A. 2001. Farmacognosia y productos naturales. Plaza de la Revolución. Cuba: Empresa Editorial Poligráfica Félix Varela.

Norma Ramal de Salud Pública # 312. 1992. Medicamentos de origen vegetal. Extractos fluidos y tinturas. Métodos de ensayo. La Habana: Editorial Pueblo y Educación.

Ocampo, R.A. 1994. Domesticación de plantas medicinales en Centro América. Colección Diversidad Biológica y Desarrollo Sustentable I. Especies Nativas. Turrialba, Costa Rica.

Parrales, C.A.A., Villamar, L.J.L. 2018. Estudio Farmacognóstico y fitoquímico preliminar de las hojas de Smilax purhampuy. BSc. diss., Universidad de Guayaquil, Facultad de Ciencias Químicas.

Rivas, Md. P.P., Muñoz, D.G.L., Ruiz, M.A.C., Fernández, L.F.T., Muñoz, F.A.C., Pérez, N.M. 2017. Global Biodiversity Information Facility (GBIF). [Online]. www.gbif.org/occurrence/search? taxon_key=2740627&advanced=1

Sen, S., Chakraborty, R., Reddy, U.Y.S.R., et al. 2010. Free radicals, Antioxidants, Diseases and Phytomedicines: Current Status and Future Prospect. Int. J. Pharm. Sci. Rev. Res. 3, no. 1: 91–100.

Silva, M.O., Brigide, P., Viva de Toledo, N.M., Canniatti-Brazaca, S.G. 2018. Phenolic compounds and antioxidant activity of two bean cultivars (Phaseolus vulgaris L.) submitted to cooking. Brazilian Journal Food Technology 21. http://dx.doi.org/10.1590/1981-6723.7216

Téllez, O. 1996. Fascículo 11: Smilacaceae Vent. 1era. Edición. 5–7 ISBN: 968-36-5726-5.

WFO. 2021. Smilax purhampuy Ruiz. Last modified February 2. www.worldfloraonline.org/taxon/wfo-0000741701

Wilson, L. 2016. Spices and Flavoring Crops: Tubers and Roots, Encyclopedia of Food and Health, 93–97. ed. Academic Press. Cambridge: Elsevier. https://doi.org/10.1016/B978-0-12-384947-2.00781-9.

World Health Organization. 1998. Quality control methods for medicinal plant materials. https://apps.who.int/iris/handle/10665/41986 (accessed Mar 7, 2021).

Zhang, X., Wu, X., Hu, Q., et al. 2019. Lab for Trauma and Surgical Infections. Mitochondrial DNA in liver inflammation and oxidative stress. Life Science 236: 116464.

11 *Malva pseudolavatera* Webb and Berthel.

Glenda Sarmiento, Yamilet Irene Gutiérrez Gaitén, Pilar Soledispa, Zoraida Burbano and Migdalia Miranda Martínez

CONTENTS

DOI: 10.1201/9781003173991-14

INTRODUCTION

The Malvaceae family, typical of the Mediterranean region, has great economic impact as it is made up of a rich diversity of species for textile, medicinal and fine ornamental use. Among the representatives of the family is the Malva genus, originally from Asia and North Africa, although it is currently distributed on all continents. It presents a wide variety of chemical constituents, such as polysaccharides, coumarins, flavonoids, polyphenols, vitamins, terpenes and tannins, among other compounds, which are found in different organs of the plant, especially in the leaves and flowers (Sharifi et al., 2019; Jedrzejczyk and Rewers, 2020). From a biological point of view, malva species have antimicrobial, anti-inflammatory, healing, strong antioxidant activity and anticancer properties (Sharifi et al., 2019; Palosch et al., 2019).

Malva pseudolavatera Webb & Berthel. is an annual or biennial subshrub that grows in fields and roadsides in coastal areas and low-altitude mountain regions. It is commonly known as "tree mallow" in North America and "khubbaza" in the Middle East (Steven, 2012). Due to its great benefits as a medicinal plant, *M. pseudolavatera* is sold in several traditional markets in Europe and America. In Spain it is used as a remedy for influenza, poor digestion, laxative, antitussive and anti-inflammatory. In Portugal it is used as an anti-inflammatory, analgesic, antiseptic, cholagogue, antiparasitic, laxative and for the treatment of wounds. In Italy it is mainly reported for gastrointestinal problems (El Khoury et al., 2020). In Ecuador the infusion is used for the treatment of eczema (De la Torre et al., 2008). Biological studies showed that *M. pseudolavatera* Webb & Berthel. aqueous leaf extracts were able to scavenge free radicals and inhibit lipoxygenase activity *in vitro*, indicating its potent antioxidant and anti-inflammatory activities (Ben-Nasr, Aazza, Mnif, and da Graca, 2015). In more recent research, the antiproliferative and pro-apoptotic effect on acute myeloid leukemia cell lines was demonstrated (El Khoury et al., 2020).

This chapter aims to provide information on the main aspects related to geographic distribution, taxonomic characteristics, micromorphological and genetic studies, as well as the main chemical constituents and biological activities demonstrated for the species *M. peudolavatera*.

GENERALITIES

In Ecuador you can find various species of Malva, one of them being *Malva pseudolavatera* Webb & Berthel., also called White Mallow. It is widely used in Ecuador for gastric problems (Balslev et al., 2008). There is no official data on its composition, its properties or its use; having a close relationship with *M. sylvestris* in terms of its medicinal use.

DEFINITION

Malva pseudolavatera Webb & Berthel. (Figure 11.1) is an annual or biennial shrub that grows in fields and roadsides in coastal areas and low-altitude mountain regions (Steven, 2012), belonging to the Malvaceae family (Edgecombe, 1964).

FIGURE 11.1 *Malva pseudolavatera* in its natural habitat, Riobamba city, Chimborazo province, Ecuador. (Picture was taken by authors during the collection of plant). (**A**) bush, (**B**) leaves, (**C**) flowers

SYNONYMS

Previously named *Lavatera cretica* (Malvaceae family), the species was transferred to the *Malva* genus and is currently called *M. pseudolavatera* or *Malva linnaei* or *Malva multiflora* (Webb and Berthelot, 1836). *Malva pseudolavatera* Webb & Berthel. is the accepted name of the species (Jepson Flora Project, 2021).

TAXONOMY

The classification obtained by the Guay Herbarium of the Faculty of Natural Sciences of the University of Guayaquil—Ecuador (Herbario GUAY):

- Class: Equisetopsidia C. Agardh
- Subclass: Magnoliidae Nóvak ex. Takht
- Superorder: Rosanae Takht
- Orden: Malvales Juss
- Family: Malvaceae Juss

- Genus: *Malva* L.
- Scientific name: *Malva cf. pseudolavatera* Webb & Berthel.

SELECTED VERNACULAR NAMES

Malva Blanca, Malva española, malva lustrada, malvilla (Balslev et al., 2008).

DESCRIPTION

It is a shrub: annual, biennial and terrestrial herbaceous. It blooms in winter, in the mountains (cold weather) from June to December. It has linear, deciduous stipules. Simple spiral-shaped leaves are broadly ovate to subdeltoid, gently lobed and crenate, 7–12 x 77–12 cm, with stellate and tomentulous trichomes on the underside, mainly on the veins, 7- palmatinervia from the base, cordate or subcordate base; petiole 6–10 cm long; bracts of involucel 3, deltoids; calyx with 5 sepals fused at the base, triangular lobes; corolla blue, white towards the base; monadel stamens; mericarpos 7, dehiscent, glabrous. Its flowers have 5 white petals. The leaves have a palmatifid, petiolate, toothed, palminervia shape, with a pubescent surface; they measure on average in their adult state 15 cm long and 20 cm wide (Steven, 2012) (Herbario GUAY).

GEOGRAPHICAL DISTRIBUTION

The plant is located on coastal bluffs, dunes, occasionally inland; elevation < 50 m. Distribution outside California: native to Southern Europe (Herbario GUAY).

The plant is distributed in Riobamba city, Chimborazo province, located in the geographic center of Ecuador, in the Andes mountains at 2,750 meters above sea level with the following coordinates: 1°40'15.5"S 78°38'49.6"O (Sarmiento et al., 2020a). It grows spontaneously in almost the entire mountain region; it requires a cold climate for its development.

GENERAL IDENTITY TESTS

Macroscopic inspection (Krapovickas, 1965; Alonso, 2007), micromorphological characteristics (Gattuso and Gattuso, 1999; Miranda and Cuellar, 2000), molecular barcode (Stecher et al., 2020, 1237), physical chemical parameters of the raw drug, phytochemical screening for the identification of secondary metabolites (Miranda and Cuellar, 2000).

ORGANOLEPTIC PROPERTIES

The leaves of *Malva pseudolavatera* have a dark green color, a characteristic smell, a fragile texture, a taste somewhat to very bitter (Krapovickas, 1965; Alonso, 2007).

MICROSCOPIC CHARACTERISTICS

The cross section at the level of the central nerve of the leaf (Figure 11.2A) displays a convex shape in its upper part and a concave shape in the lower part; the epidermis of these faces is provided with few star-shaped trichomes. An epidermis with a cellular stratum is observed, followed by the collenchyma made up of compact cells. Later, the spongy parenchyma is presented and the vascular bundles (xylem and phloem) are more in the center (Sarmiento et al., 2020a).

In an image of the lateral arm of the mesophyll (Figure 11.2B) trichomes are observed; their grouping is 2 to 3 pieces. On both surfaces (adaxial and abaxial) the epidermis made up of tabular cells can be seen, most notably on the upper one, followed by palisade tissue. A palisade tissue and spongy parenchyma are also found (Sarmiento et al., 2020a).

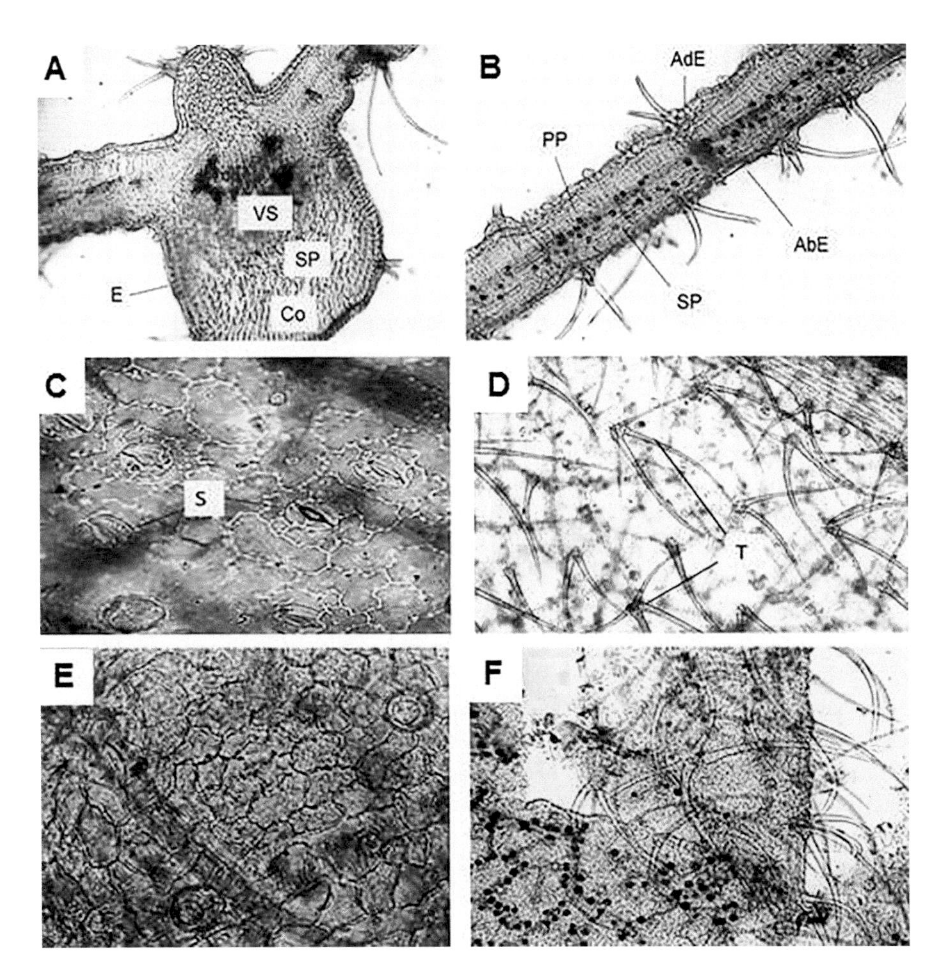

FIGURE 11.2 Micromorphology of *Malva pseudolavatera* leaves- Transversal section of the central nerve of the leaf and diaphanized: (A) Central nerve. (B) Mesophyll. (C) abaxial epidermis and stomata. (D) Trichomes. E: epidermis, Co: collenchyma, SP: spongy parenchyma, VS: vascular system, AbE: abaxial epidermis, PP: palisade parenchyma, Ade: adaxial epidermis, S: stomata, T: trichomes. Micro-morphological details of the powder drug. (E) Epidermal cells. (F) Epidermal cells and trichomes

In the clearing of the abaxial epidermis (Figure 11.2C) of the leaf, slightly sinuous cells with stomata of the anisocytic or cruciferous type were visualized (three attached cells and one smaller one). The enlarged image of the trichomes (Figure 11.2D) will show the grouping of these structures with a semi-starry shape where their point of attachment is not pigmented (Sarmiento et al., 2020a, 3554).

Analysis of the powdered drug revealed slightly sinuous polygonal epidermal cells (Figure 11.2E). The conforming elements of the semi-stellate structure of the hairs were also observed, as well as the dots or black structures (Figure 11.2F) (Sarmiento et al., 2020a).

MOLECULAR BARCODE

Sequences were trimmed from low quality using Chromas (Technelysium). Publicly available barcode sequences from the Malvoideae were queried in the GenBank database (29th May 2020) and used in phylogenetic analysis using MEGAX (Stecher et al., 2020). Furthermore, for the *rbc*L and *mat*K barcode sequence, the recommended model after analysis of the queried sequences using MEGAX was used after alignment with MUSCLE. For the phylogenetic analyses, the Maximum Likelihood method was used for each barcode using bootstrap test (100 replicates). To compare the analysis with different genera from the Malvoideae subfamily, 3 accessions from the *Alcea* and the *Althae* genera were used in the phylogenetic analysis for each barcode.

The analysis revealed that for the rbcL barcode, in the majority of *Malva* spp. the sequences are grouped in different clades; a similar behavior was observed in matK. Meanwhile, the concatenated sequence of the ITS (ITS1 and ITS2) presented the formation of different clades in specific *Malva* species. *M. pseudolavatera* from Ecuador (accession MH513499) was grouped with several accessions of *M. dendromorpha* (EF419468, EF419467, EF419466, EF419469, AF303020) with a bootstrap value of 100. A total of 26 variable sites were found in an alignment of 872 nucleotides of the concatenated ITS region (ITS1 and ITS2). In the same alignment, a space was observed at position 325 for *M. pseudolavatera*. The ITS region was more informative than rbcL and matK to discriminate different species. With the markers rbcL and matK no differentiation was found in the phylogenetic analysis, but when performing the analysis with the ITS; the result of the Blast-N (sequences from the NCBI database) with the markers rbcL, matK, its1, its2, it was possible to show that the sample corresponds to the species of the genus *Malva* spp., because the ITS markers are very useful for inter-species analysis. The genetic study has not been referred to previously and allowed corroborating the taxonomic classification of the cultivated species in Ecuador (Sarmiento et al., 2020a).

PURITY TESTS

Physical-Chemical Parameters of Raw Drug

- ***Moisture content:*** 10.14/0.14% (mean/standard deviation) (Sarmiento et al., 2020b), between 8–14% (Zhi-cen, 1980; WHO, 2011).

- ***Total ash***: It was 15.25/0.14% (Sarmiento et al., 2020b); Chinese Pharmacopoeia refers up to 15% (Commission CP, 2015).
- ***Water-soluble ash***: 4.17/0.06% (Sarmiento et al., 2020b).
- ***Acid-insoluble ash***: 13.02/0.08% (Sarmiento et al., 2020b).
- ***Water-soluble extractive***: 27.38/1.01% (Sarmiento et al., 2020b).
- ***Alcohol-soluble extractive at 30%***: 19.51/0.59% (Sarmiento et al., 2020b).
- ***Alcohol-soluble extractive at 50%***: 17.10/0.10% (Sarmiento et al., 2020b).
- ***Alcohol-soluble extractive at 80%***: 10.92/0.32% (Sarmiento et al., 2020b).
- ***Alcohol-soluble extractive at 98%***: 7.39/0.33% (Sarmiento et al., 2020b).

Physical-chemical parameters of the extract (using an 80% hydroalcoholic mixture as solvent) (Miranda and Cuellar, 2000).

- ***pH***: 5.83/0.02
- ***Total solids***: 1.40/0.05%
- ***Refraction index***: 1.3593/0.0002
- ***Relative density:*** 0.8743/0.0015 mg/L
- ***Capillary analysis*** with height of 8.1/0.14 cm.

CHEMICAL ASSAYS

Phytochemical screening (Miranda and Cuellar, 2000); total phenols by the Folin-Ciocalteu method (Pourmorad et al., 2006; Chlopicka et al., 2012) and total flavonoids by the colorimetric method of aluminum chloride (Pourmorad et al., 2006; Chang et al., 2002); thin layer chromatography (TLC), gas chromatographic-mass spectrometry (GC-MS), high performance liquid chromatography- mass spectrometry (HPLC-MS).

Phytochemical screening at raw drug: Fats or oil, alkaloids, triterpenoids, mucilages, saponins, amino acids, reducing substances, tannins/phenols, quinones, flavonoids, anthocyanidins, catechins (Sarmiento et al., 2020b).

Total phenols: 1.47/0.04 mg/mL (Sarmiento et al., 2020b).
Total flavonoids: 1.22/0.01mg/mL (Sarmiento et al., 2020b).
Thin layer chromatography (TLC): it was observed that some spots had a characteristic behavior of phenolic and triterpenoids structures.

Gas Chromatographic-Mass Spectrometry (GC-MS)

Saponifiable fraction (major chemical constituents): Hexadecanoic acid (palmitic acid 27.42/7.65%), 9,12-octadecadienoic acid (linoleic acid 11.02/3.46%) and 9,12,15-octadecatrienoic acid (linolenic acid 61.56/6.75%) (Sarmiento et al., 2020b).

Unsaponifiable fraction (major chemical constituents): neophytadiene, phytol, tricosane, tetracosane, pentacosane, heptacosane, octacosane, squalene, nonacosane, Stigmasta-5-en-3β-ol and nonacosane (42.99/1.17%) (Sarmiento et al., 2020b).

Hydroalcoholic extract: glucitol (14.91%), α-linoleic acid (13.75%), palmitic acid (13.40%), stearic acid (14.98%) and octadecanoic acid (14.98%) (Sarmiento et al., 2020b).

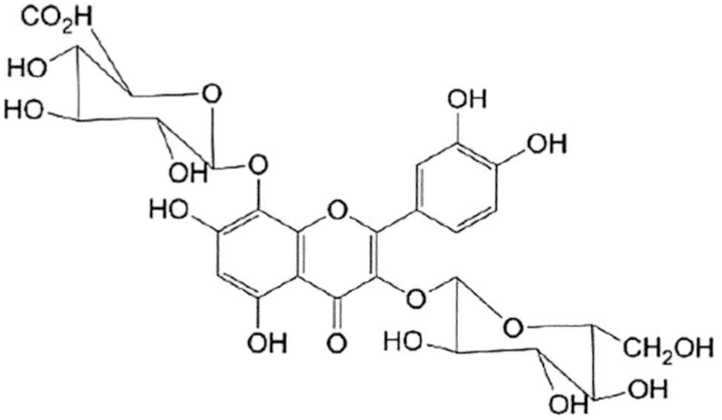

FIGURE 11.3 Structure of gosipetin 3-***O*** -***β*** -D-glucopyranosyl-8-***O*** -***β*** -D-glucuronopyranoside

High Performance Liquid Chromatography-Mass Spectrometry (HPLC-MS)

Major chemical constituents: gosipetin 3-*O*-*β*-D-glucopyranosyl-8-*O*-*β*-D-glucuronopyranoside (is identified for the first time for the species) (Figure 11.3); malic acid, hydroxybenzoic acid, protocatechic acid, 4-methoxybenzoic acid, 4-hydroxydihydrocinnamic acid, hydroxybenzoic acid and rhamnosylglucosylflavonoid (Sarmiento et al., 2020c.)

MEDICINAL USES

Among the 50 species of Malva, *M. pseudolavatera* Webb & Berthel. is known for its versatile uses (Singorini et al., 2009). In Spain, it is considered as a remedy for influenza, upper respiratory tract infections and cough (Rivera et al., 1993), whereas in Portugal it is used for its laxative, analgesic and antiseptic effects (Rivera et al., 1993; Balslev et al., 2008).

PHARMACOLOGY

Experimental Pharmacology Gastroprotective Activity

Aqueous decoction extracts were made which were lyophilized to obtain dry extracts. Gastroprotective activity was studied in two models of ulcers induced by ethanol (Coronel, 2016) and NSAIDs (ASA) (Huaman et al., 2009).

Male mice with average weights of 30 ± 2 g were used. The treatments of aqueous extracts of *M. pseudolavatera* were administered orally, for seven days, in doses of 1,000, 500, 250, 125 and 75 mg/kg of mouse weight. It was evidenced that at the dose of 1,000 mg/kg, *M. pseudolavatera* presented a high gastric protection, which was met at lower doses for the two models used. The inhibition percentages of gastric ulcers showed that the positive control treated with Sucralfate reached 87.5% inhibition, while the group treated with the extracts of *M. pseudolavatera* at doses of

1,000 mg/kg reached 77% inhibition in the induction model alcohol and 84% inhibition in the induction model NSAID (Sarmiento et al., 2020d).

Antioxidant Activity

The antioxidant capacity was performed by the Ferric Reducing Antioxidant Power (Benzie and Strain, 1996), 2,2-diphenyl-1-picrylhydrazyl (Kedare and Singh, 2011) and 2,2'-azino-bis(3-ethylbenzothiazoline)-6-sulfonic acid (Re et al., 1999, 1231; Re et al., 1999) techniques. In the FRAP test, antioxidant activity was evidenced in a concentration-dependent manner. There was a tendency to increase the antioxidant capacity by FRAP as the concentration of the extract increased (Sarmiento et al., 2020c).

In the DPPH test it was observed that as the concentration of the extract increased, the inhibition of the radical increased. An important aspect to consider is the determination of the IC_{50}. In this sense, the extract showed good anti-radical activity (IC_{50}= 82.53/7.18 µg/mL), although the highest activity was for vitamin C (IC_{50} = 52.73/9.15 µg/mL) (Sarmiento et al., 2020c).

For the ABTS test, there was a tendency to increase the inhibition capacity of the radical as the concentration increases. For the percentages of inhibition of the radical ABTS, significant differences were observed between the samples tested. The extract showed at the minimum concentration (100 µg/mL) a sequestration capacity greater than 50%, even greater, at the same concentration evaluated for the two reference substances (trolox and vitamin C). At concentrations of 200 and 300 µg/mL, the extract exhibited a higher percentage of inhibition than Trolox, which suggests a high antioxidant activity. Of the evaluated samples, the one that presented the lowest IC_{50} and therefore the highest antioxidant activity was vitamin C (196.20/5.26 µg/mL). However, *M. pseudolavatera* extract was able to inhibit 50% (IC_{50}= 290.00/7.49 µg/mL), also showing anti-radical activity (Sarmiento et al., 2020c).

Hypoglycemic Activity

The method described by Pinzon *et al.* (1995) was followed for this trial. The hypoglycemic activity in Wistar rats at the doses of 250, 500 and 1,000 mg/kg, of the extracts, were evaluated. The positive control group treated with the hypoglycemic drug metformin before and after induction with alloxan, presented a blood glucose concentration of 119.67 ± 43.50 mg/dL, because this drug increases the basal rate of the transport of glucose. The negative control group presented a significant difference with the rest of the groups, with an average glucose value of 455 ± 113.86 mg/dL (Sarmiento et al., 2020c).

For the groups treated with the extracts of *M. pseudolavatera*, the glucose values were 124.33 ± 7.50 mg/dL at dose of 1,000 mg/kg, 125.40 ± 21.73 mg/dL at dose of 500 mg/kg and 129.40 ± 32.84 mg/dL at dose of 250 mg/kg; there was no significant difference between them. This study demonstrated that the extract of the species *M. pseudolavatera* has a high normoglycemic effect Sarmiento et al., 2020c).

Mucolytic Activity

The precipitate of the aqueous extracts of *M. pseudolavatera* (mucilage) and the phenol red method (Engler and Szelenyi, 1984) were used. The administration of the precipitate and bromhexine was carried out orally and phenol red intraperitoneally in male mice. Different doses of the precipitate were (250, 500 and 1,000 mg/kg). After 24 hours of animal fasting, the administration of the 300 mg/kg phenol red indicator was administered to all groups of animals by the intraperitoneal route, except for the control group (-). An hour after the administration of phenol red, all groups of animals was sacrificed, followed by the removal of their tracheas, for their respective evaluation (Sarmiento et al., 2020b).

It is possible to show the highest percentages of activity in the higher doses, which indicates that there is a directly proportional relationship, dose-response; the higher the dose, the greater the effect; presenting a better percentage of efficacy for *M. pseudolavatera* in the doses of 500 and 1,000 mg/kg against the reference drug bromhexine (Sarmiento et al., 2020b).

Anti-cancer Properties

Treatment of myeloid leukemia (AML) cell lines with *M. pseudolavatera* methanolic leaf extract showed a dose- and time-dependent inhibition of proliferation and a dose-dependent increase in apoptotic hallmarks such as an increase in phosphatidylserine on the outer membrane leaflet and membrane leakage in addition to DNA fragmentation. The pro-apoptotic effect was induced by reactive oxygen species (ROS) as well as an upregulation of cleaved poly (ADP-ribose) polymerase (PARP), increase in Bax/Bcl-2 ratio, andrelease of cytochrome-c from the mitochondria (El Khoury et al., 2020).

Acute Oral Toxicity

At the dose of 2,000 mg/kg of body weight of the experimental animals (Wistar rats), no clinical signs were recorded, there was no mortality and no significant differences were found between the body weights of the different groups, nor food and water consumption. Histopathology was not performed because there was no alteration in organs. The aqueous extract of *M. pseudolavatera* at the dose evaluated shows a wide safety range under the test conditions detailed according to OECD 423 (2001).

CLINICAL PHARMACOLOGY

None reported scientifically for this species.

WARNINGS

No warnings have been reported.

CONCLUSIONS

M. pseutolavatera is a very widespread and commonly used species in Ecuador. Its identity and purity was demonstrated through pharmacognostic studies, with micromorphological characterization and DNA barcode analysis being transcendental. A varied composition in metabolites was detected that is related to the pharmacological activities demonstrated for the species, as well as the non-acute oral toxicity of aqueous extracts of the plant. Specific data on tests on genotoxicity, reproductive toxicity and carcinogenicity have not been found, nor clinical studies. Further chemical and biological studies are required to evaluate other activities related to its traditional use, especially those that have already been demonstrated in other species. *M. pseudolavatera* undoubtedly constitutes a promising species with high therapeutic potential that can be considered in the preparation of phytomedicines. The data provided in this chapter constitute fundamental aspects in the development of the plant monograph and contribute to its scientific validation as an herbal product, which has been used for a long time by the Ecuadorian population.

REFERENCES

Alonso, J. 2007. Malva. Tratado de Fitofármaco y Nutracéutico, 2 ed. Editorial Corpus: Rosario Argentina; 671.

Balslev, H.; Navarrete, H.; De la Torre, L.; Macía, M. 2008. Introducción, in Balslev (eds.) *Enciclopedia de las Plantas Útiles del Ecuador*, Quito & Aarhus: Herbario QCA & Herbario AAU, 1–3.

Ben-Nasr, S.; Aazza, S.; Mnif, W.; da Graca C. M. M. 2015. Antioxidant and anti-lipoxygenase activities of extracts from different parts of Lavatera cretica L. grown in Algarve (Portugal). Pharm. Mag. 11: 48–54.

Benzie, I. F. F.; Strain, J. J. 1996. The ferric reducing ability of plasma (FRAP) as a measure of antioxidant power: The FRAP assay. Anal. Biochem. 239(1): 70–76.

Chang, C.C.; Yang, M.H.; Wen, H.M.; Chern, J.C. 2002. Estimation of total flavonoid content in propolis by two complementary colorimetric methods. J Food Drug Anal 10: 178–182. doi: 10.38212/2224–6614.2748

Chlopicka, J.; Pasko, P.; Gorinstein, S.; Jedryas, A.; Zagrodzki, P. 2012. Total phenolic and total flavonoid content, antioxidant activity, and sensory evaluation of pseudocereal bread. LWT- Food Sci Technol 46: 548–555. doi: 10.1016/j.lwt.2011.11.009

Commission, C.P. 2015. Pharmacopoeia of the People's Republic of China. Peking: Chinese Medical Science and Technology Press; 337.

Coronel, E. 2016. Efecto regenerador del extracto acuoso de semilla de *Linum usitatissimum* (linaza) sobre la mucosa gástrica con úlcera inducida por etanol en ratas. Tesis. Lima: Universidad nacional mayor de San Marcos, Facultad de Medicina. Available from: https://pdfs.semanticscholar.org/957f/7df3ea9b2e16229666d72ebd324084dd55f5.pdf?_ga=2.152230393.1977122881.1596918115-381412144.1589480874

Edgecombe, W.S. 1964. Weeds of Lebanon. Beirut, Lebanon: American University of Beirut; 244–247.

El Khoury, M.; Haykal, T.; Hodroj, M. H.; Abou, N. S.; Sarkis, R.; Taleb, R. I; et al. 2020. Malva pseudolavatera leaf extract promotes ros induction leading to apoptosis in Acute Myeloid. Leukemia Cells In Vitro Cancers, 12, 435. doi: 10.3390/cancers12020435

Engler, H.; Szelenyi, I. 1984. Tracheal phenol red secretion, a new method for screening mucosecretolytic compounds. J Pharmacol Methods 11: 151–157. doi: 10.1016/0160–5402(84)90033-0

Gattuso M.A.; Gattuso S.J. 1999. Manual de procedimientos para el análisis de drogas en polvo. Editorial de la Universidad Nacional de Rosario Urquiza. Argentina. ISBN N◦ 950-673-199-3.

Herbario GUAY. Facultad de Ciencias Naturales de la Universidad de Guayaquil. https://docplayer.es/63819333-Herbario-guay-facultad-de-ciencias-naturales-universidad-de-guayaquil-casilla-guayaquil-ecuador-b.html

Huamán, O; Sandoval, M; Arnao, I; Béjar, E. 2009. Antiulcer effect of lyophilized hydroalcoholic extract of *Bixa orellana* (annatto) leaves in rats. An. Fac. Med. 70(2). Available from: www.scielo.org.pe/scielo.php?script=sci_arttext&pid=S1025-55832009000200003

Jedrzejczyk, I.; Rewers, M. 2020. Identification and genetic diversity analysis of edible and medicinal Malva species using flow cytometry and issr molecular markers. Agronomy 10, 650. doi: 10.3390/agronomy10050650

Jepson Flora Project (eds.) 2021, *Jepson eFlora*. https://ucjeps.berkeley.edu/eflora/, accessed on April 15, 2021.

Kedare, S.B.; Singh, R.P. 2011.Genesis and development of DPPH method of antioxidant assay. J Food Sci Technol. 48(4): 412–422.

Krapovickas, A. 1965. Notas sobre Malvaceae III. *Kurtziana* 2: 113–126.

Miranda, M.M.; Cuellar, A.C. 2000. Laboratory practice manual. *Pharmacognosy and natural products*. Havana: Editorial Felix Varela, 25–49, 74–79.

OECD 423. 2001. OECD Guideline for testing of chemicals. Acute Oral Toxicity—Acute Toxic Class Method.

Paloschi, de Oliveira L.; Bovini, M. G.; Lopes, da Costa B. R.; Carissimi, B. M. I. & Boff, P. 2019. Species of Malva L. (Malvaceae) cultivated in the western of Santa Catarina state and conformity with species marketed as medicinal plants in southern Brazil. Journal of Agricultural Science 11(15). doi: 10.5539/jas.v11n15p171

Pinzón, R.; Gupta, M.; San Román, L. 1995. Búsqueda de Principios Bioactivos en Plantas de la Región. Manual de Técnicas de Investigación. Colombia. CYTED. Subprograma X Proyecto X-1., 173–182.

Pourmorad F.; Hosseinimehr S.J.; Shahabimajd N. 2006. Antioxidant activity, phenol and flavonoid contents of some selected Iranian medicinal plants. Afr J Biotechnol 5: 1142–1145.

Re, R.; Pellegrini, N.; Proteggente, A.; Pannala, A.; Yang, M.; Rice, E. C. 1999. Antioxidant activity applying an improved ABTS radical cation decolorization assay. Free Radic Biol Med. 26: 1231–1237.

Rivera, N.D.; Obon, D.C.C. 1993. Medicaments et Aliments: l'approche ethnopharmacologique. Acte du 2eme Colloque Europeen d'Ethnopharamcologie et de la 1ere Coference Internationale d'Ethnomedicine. Heidelberg 2: 223.

Sarmiento, T.G.; Santos, E.O.; Miranda M.M.; Pacheco, C. R.; Scull, L. R.; Gutiérrez, G.Y.; Delgado, H. R. 2020a. Molecular barcode and morphology analysis of *Malva pseudolavatera* Webb & Berthel and *Malva sylvestris* L. from Ecuador. Biodiversitas 21(8): 3554–3561. doi: 10.13057/biodiv/d210818

Sarmiento, T.G.M.; Miranda, M. M.; Chóez, G. I. A.; Gutiérrez, G. Y. I., Delgado, R. H; Carrillo, L. G. 2020b. Pharmacognostic, chemical and mucolytic activity study of *Malva pseudolavatera* Webb & Berthel. and *Malva sylvestris* L. (Malvaceae) leaf extracts, grown in Ecuador. Biodiversitas 21(10): 4755–4763. doi: 10.13057/biodiv/d211040

Sarmiento, T.G.M.; Miranda, M. M.; Gutiérrez, G. Y. I.; Delgado, H. R.; Carrillo, L.G. 2020c. Chemical Study, Antioxidant Capacity, and Hypoglycemic Activity of *Malva pseudolavatera* Webb & Berthel and *Malva sylvestris* L. (*Malvaceae*), Grown in Ecuador. Trop J Nat Prod Res, 4(12): 1064–1071. doi: 10.26538/tjnpr/v4i12.7

Sarmiento, T.G.M.; Delgado, H. R.; Carrillo, L. G.; Miranda M. M. 2020d. Comparative gastroprotective activity of *Malva pseudolavatera* Webb & Berthel and *Malva sylvestris L.* grown in Ecuador. Pharmacologyonline 2: 1–12.

Singorini, M.A.; Piredda, M.; Bruschi, P. 2009.Plants and traditional knowledge: An ethnobotanical investigation on Monte Ortobene. J. Ethnobiol. Ethnomed. 5, 6.

Sharifi, J. R.; Melgar, L. G.; Hernández, A. A. J.; Taheri, Y.; Shaheen, S.; Kregiel, D.; et al. 2019. Malva species: Insights on its chemical composition towards pharmacological applications. Phytotherapy Research, 1–22. doi: 10.1002/ptr.6550

Stecher G.; Tamura K.; Kumar S. 2020. Molecular Evolutionary Genetics Analysis (MEGA) for macOS. Mol Biol Evol 37 (4): 1237–1239. doi: 10.1093/molbev/msz312

Steven, R.H. 2012. *Malva pseudolavatera*, in Jepson Flora Project (eds.) *Jepson e Flora*, https://ucjeps.berkeley.edu/eflora/eflora_display.php?tid=89042, accessed on April 15, 2021.

Webb, P.B.; Berthelot, S. 1836. *Histoire Naturelle des Iles Canaries, Phytographia Canariensis*; Bethune, editeur: Paris, France, 1836–1850.

WHO (World Health Organization). 2011. Quality control methods for medicinal plant materials. WHO/PHARM/92.559. Updated edition of Quality control methods for medicinal plant materials, WHO, Geneva.

Zhi-cen, L. 1980. General control methods for vegetable drugs. Comparative study of methods included in thirteen pharmacopoeias a proposal on their international unification. www. WHO/PHARM/80.502. 8–39.

12 Nutraceutical and Functional Properties of the Andean Grain *Amaranthus caudatus* L.

Adriana Orellana-Paucar

CONTENTS

DOI: 10.1201/9781003173991-15

INTRODUCTION

A nutraceutical is a naturally occurring food compound marketed as a dietary supplement in a dosage form. This compound may possess beneficial health effects when administrated in higher quantities than those present in food. Moreover, when consumed as part of the diet in recommended amounts for daily intake, a functional food provides an adequate nutritional effect and contributes to health improvement or reduces the risk of disease (Luengo 2007). In this context, due to its characteristic protein composition with a unique qualitative and quantitative protein profile and its appealing pharmacological properties, *Amaranthus caudatus* is currently considered an appealing nutraceutical, functional food and drug candidate. Worldwide interest in amaranth (*A. caudatus*) began in the 1980s, mainly favored by a recognition endorsed by the U.S. National Academy of Sciences. Then, amaranth was included among the world's most promising crops due to its nutritional value (National Research Council 1984).

BOTANICAL AND PHYSIOLOGICAL DESCRIPTION

Amaranthus caudatus L. belongs to the Amaranthaceae botanical family. It is commonly known in the Ecuadorian Highlands as "kiwicha", "sangorache" or "ataco". Amaranth is a broad-leafed pseudocereal with purple inflorescences, stems and leaves. Inflorescences could reach up to 50 cm long. A botanical specimen of *A. caudatus* is depicted in Figure 12.1. Amaranth grains are pale cream-colored, with diameters between 0.9- and 1.7-mm. It contains approximately 50,000 seeds per plant (National Research Council 1984). Figure 12.2 portrays a sample of harvested amaranth grains.

Amaranthaceae family belongs to the group of plants with C4 pathway photosynthesis. The C4 route leads to a better conversion of atmospheric carbon into plant sugar than the classical C3 cycle. This group of plants can also photosynthesize even under adverse climatic conditions such as high temperature, salinity, or drought (National Research Council 1984).

ORIGIN AND GEOGRAPHICAL DISTRIBUTION

The Amaranthus genus is widely distributed to temperate and even tropical or subtropical regions. This genus possesses about 70 species. Most of them are native to America and approximately 15 to Europe, Asia, Africa and Oceania. Amaranthus species are classified into 4 groups according to their use: grain, vegetable, ornamental and weedy amaranth. The grain amaranth group comprises *A. caudatus, A. hypochondriacus, A. cruentus* and *A. edulis* (Martinez-Lopez et al. 2020). *Amaranthus caudatus* is native to Ecuador, Peru and Bolivia. It commonly grows above 2,500

FIGURE 12.1 *Amaranthus caudatus*. (Author: Matias Villacís-Luzuriaga. Cuenca, Ecuador.)

FIGURE 12.2 *Amaranthus caudatus* seeds. (Author: Matias Villacís-Luzuriaga. Cuenca, Ecuador.)

meters of altitude. Other species such as *A. cruentus* L. and *A. hypochondriacus* L. are typically cultivated in Central America. Amaranth plants resist pests, drought, heat and freezing. *A. caudatus* resists freezing more effectively than other species of this botanical family (National Research Council 1984, 16–27). Amaranth constituted a staple food for Incas (Ecuador, Peru and Bolivia), Mayas and Aztecs (Mexico). South and Central American natives mixed amaranth grains with honey or human blood to shape them into assorted shapes (i.e., birds, mountains, animals) to be eaten during ceremonies or social gatherings. This custom seemed to scandalize Spanish colonizers, who strongly prohibited amaranth farming (National Research Council 1984). The Spanish invasion restrained the further development of the native cultures in many aspects, including amaranth's nutritional and therapeutic application.

NUTRITIONAL COMPOSITION

MACRONUTRIENTS

A. caudatus is a prominent source of proteins, carbohydrates, dietary fiber, lipids, vitamins and minerals (Amico and Schoenlechner 2017; Herrera Fontana et al. 2020; Soriano-García and Aguirre-Díaz 2020). Amaranth seeds contain a higher concentration of proteins, fat and dietary fiber than flour. Table 12.1 depicts the nutritional composition of amaranth.

Amaranth starch shows improved freeze-thaw stability and higher gelatinization temperature and viscosity compared to corn. The mono- and disaccharides content in amaranth is minimal (3–5%). Starch is the major component of amaranth seeds. Total dietary fiber (soluble and insoluble) in amaranth seeds (9.3%) is more significant than in rice (0.9%), maize (2.3%) and wheat (2.6%) (Arendt and Zannini 2013). Approximately 75% of amaranth oil composition comprises polyunsaturated fatty acids. The major constituents of the oil are palmitic and oleic acids and the essential fatty acid: linolenic (Arendt and Zannini 2013). The protein composition of *A. caudatus* grains includes a total content of essential amino acids in a similar amount to the recommended dietary allowances (RDA) suggested by the Food and Agriculture Organization (FAO), the World Health Organization (WHO) and the United Nations University (UNU) for adults (Food and Agriculture Organization, World Health Organization, and United Nations University 2007).

TABLE 12.1
Macronutrient Content of *Amaranthus caudatus* Grains and Flour

Component	Amaranth grain (g per 100 g)	Amaranth flour (g per 100 g)
Energy (kcal)	381.0	390.1
Proteins	14.4	12.0
Total fat	6.5	6.1
Carbohydrates	66.1	71.8
Dietary fiber	9.3	3.5

TABLE 12.2
Amino Acid Content of *Amaranthus caudatus* Grains Compared to Cereals of Daily Consumption

Amino acid	*A. caudatus*, raw grains (g/100 g protein)	Quinoa (g/100 g protein)	Wheat (g/100 g protein)	Maize (g/100 g protein)	Brown rice (g/100 g protein)	Oat (g/100 g protein)	FAO/WHO/UNU RDA (g/100 g protein)
Isoleucine	3.6–5.2	4.9	3.4	2.3–3.4	4.0	3.8–4.1	3.0
Leucine	5.7–6.4	6.6	6.7	7.9–11.4	7.9	6.9–7.6	5.9
Lysine	4.8–6.7	6.0	2.8	2.3–3.1	3.6	3.5–4.1	4.5
Sulfur amino acids	4.5–4.9	2.0–4.8	3.6	3.7–5.0	3.3	4.1–5.8	2.5
Aromatic amino acids	7.0–8.1	6.2–7.5	6.3	5.6–6.5	8.5	7.7–9.0	6.3
Threonine	3.3–5.1	3.7	2.8	2.4–2.7	3.5	3.1–3.4	2.3
Tryptophan	1.1–1.8	0.9	1.2	0.6	1.2	1.1	0.6
Valine	4.5–4.7	4.5	4.2	3.3–5.2	5.6	5.2–5.8	3.9
Histidine	1.5–1.7	3.2	2.4	2.2–3.6	2.4	2.1–2.4	1.5

TABLE 12.3
Micronutrient Content of *Amaranthus caudatus* Seeds

Minerals	Amaranth grain (value per 100 g)
Potassium	508.0–892.0 mg
Phosphorus	119.0–779.0 mg
Magnesium	279.2–427.0 mg
Calcium	147.0–283.1 mg
Iron	4.6–29.3 mg
Zinc	1.6–9.3 mg
Manganese	3.3–5.4 mg
Sodium	1.9–4.1 mg
Copper	0.5–1.2 mg
Selenium	18.7–50.0 µg
Vitamins	
Ascorbic acid	4.0–4.2 mg
Pantothenic acid	1.4 mg
Alpha-tocopherol	1.1 mg
Beta-tocopherol	0.9 mg
Delta-tocopherol	0.6 mg
Niacin	0.9 mg
Vitamin B6	0.5 mg
Riboflavin	0.2 mg
Thiamin	0.1 mg
Folate	49.0–82.0 µg
Vitamin A	2.0 IU

Table 12.2 compares RDA with the amino acid content of *A. caudatus* grains, quinoa, wheat, maize, brown rice and oat (Pomeranz, Youngs, and Robbins 1973; Shoup, Pomeranz, and Deyoe 1966: Vega-Gálvez et al. 2010; Arendt and Zannini 2013; Ahenkora et al. 2016; Juliano 2016; Youssef et al. 2016). It is noted that leucine, lysine and aromatic amino acids are the major constituents of amaranth grains. Lysine is a limiting amino acid in wheat, maize, rice and oat. Thereby, the high content of lysine and all essential amino acids in amaranth appoint it as an outstanding protein source. Furthermore, the gluten-free nature of amaranth and the high solubility and digestibility of proteins compared with most cereals confer this food an added nutritional value.

MICRONUTRIENTS

Amaranth grains constitute a valuable source of ascorbic acid, niacin, pantothenic acid and tocopherols. They also possess a noticeable content of minerals including potassium, phosphorus, magnesium, calcium, iron and zinc (Arendt and Zannini 2013; Soriano-García and Aguirre-Díaz 2020). Table 12.3 shows the detailed nutritional composition of amaranth.

An *in vivo* analysis revealed increased amaranth calcium bioavailability in rats when grains were processed before consumption. Hence, extruded amaranth grains appear to be an appealing source of dietary calcium (Ferreira and Arêas 2010).

The presence of phytates in the composition of amaranth may interfere with iron and calcium absorption at the intestinal level. However, adding up to 20% amaranth flour to wheat-based bread dough enhances its nutritional value due to a better iron bioavailability. Furthermore, tocopherols provide antioxidant protection to amaranth oil and its constituents and positively contribute to human health (Arendt and Zannini 2013).

TRADITIONAL MEDICINE USES

The infusion of stems and inflorescences of amaranth is orally administered for treating metrorrhagia and after childbirth as a laxative (Ríos et al. 2007; Molina Vélez 2008).

BIOACTIVE COMPOUNDS AND POTENTIAL NUTRACEUTICAL APPLICATIONS

The major secondary metabolites of amaranth grains are phenolic compounds, including tannins, flavonoids and phenolic acids (Tang and Tsao 2017). Triterpenoid saponins, phytosterols and betalains are also present (Martínez-Villaluenga, Peñas and Hernández-Ledesma 2020). In addition to flavonoids, the leaf extracts of amaranth contain steroids, terpenoids and cardiac glycosides (Emekli, Kendirli, and Kurunc 2012).

ANTICARCINOGENIC ACTIVITY

A. caudatus properties were assessed using the bone marrow micronucleus assay. Amaranth can decrease micronuclei formation in rats exposed to sodium arsenite. In addition, the production of gamma-glutamyl transferase and alkaline phosphatase was not affected. This property is attributed to amaranth's polyphenolic compounds, vitamins and minerals (Adewale and Olorunju 2013).

ANTIOXIDANT ACTIVITY

The main flavonoids of amaranth are rutin, isoquercetin and nicotiflorin. The antioxidant assays demonstrated significant antioxidant activity during two stages of botanical development: a) vegetative and b) flowering and grain fill. Consequently, consuming amaranth during these stages could be suggested to seize its properties (Karamać et al. 2019). In addition to the well-known antioxidant activity of rutin, it also exhibits a potential application for treating cancer, diabetes, neurodegenerative diseases, cognitive deficit, inflammation, hyperuricemia, hypertension and hypercholesterolemia (Al-Dhabi et al. 2015).

ANTI-INFLAMMATORY AND ANTIPYRETIC ACTIVITY

Saponins are responsible for the bitter taste of pseudocereals (i.e., quinoa and amaranth) and can cause toxicity in humans. Although saponins possess low cell membrane permeability and poor bioavailability, it is recommended to immerse the seeds in water before consumption to remove them from amaranth.

The application of microbial, chemical and enzymatic hydrolysis leads to the transformation of saponins to sapogenins. Sapogenins exhibit improved bioavailability and therapeutic effects in murine models of acute inflammation (Martínez-Villaluenga, Peñas, and Hernández-Ledesma 2020). Moreover, in a yeast-induced fever murine model, the methanolic extract of *A. caudatus* leaves demonstrated a potent antipyretic action, comparable with paracetamol (Peter and Gandhi 2017).

ANTINOCICEPTIVE POTENTIAL

Extract of the whole plant of *A. caudatus* exhibited a potent central and peripheral antinociceptive activity in murine models by delaying reaction time to thermal stimulation (Srinivas et al. 2010). This effect is attributed to flavonoids, terpenoids, alkaloids, glycosides, steroids and phenolic compounds present in the extract (B. Ashok Kumar et al. 2010b).

ANTIDEPRESSANT ACTIVITY

The methanolic extract of *A. caudatus* shows activity in murine models of forced swimming and tail suspension. The observed activity was comparable with the effects of escitalopram and imipramine in these models. These findings are probably related to flavonoids and polyphenolic compounds in the extract (B. Ashok Kumar et al. 2022).

ANTIBACTERIAL ACTIVITY

The hexane, ethyl acetate, dichloromethane and methanol leaf extracts of *A. caudatus* displayed significant activity against *Staphylococcus aureus, Bacillus spp.,*

Escherichia coli, Salmonella typhi, Pseudomona aeruginosae, Proteus mirabilis and *Klebsiella pneumoniae*. Flavonoids, steroids, terpenoids and cardiac glycosides appear to be responsible for this activity (Emekli, Kendirli, and Kurunc 2012).

ANTIMALARIAL ACTIVITY

A. caudatus is one of the constituents of a polyherbal medicine used to treat malaria in Nigeria. The effectiveness of this herbal antimalarial preparation was evaluated and confirmed in mouse models. The reported effect appears to be higher than the one achieved with chloroquine. Nevertheless, to date, there is no clinical confirmation of this pharmacological property (Obidike, Amodu, and Emeje 2015).

ANTIFUNGAL ACTIVITY

Ac-AMP1 and Ac-AMP2, two peptides isolated from *A. caudatus* grains, showed remarkable activity against the pathogenic fungi: *Alternaria brassicola, Ascochyta pisi, Botrytis cinerea, Colletotrichum lindemuthianum, Fusarium culmorum, Trichoderma hamatum* and *Verticillium dahliae*, at remarkably lower doses than other known antifungal chitin-binding proteins (Broekaert et al. 1992).

ANTHELMINTIC ACTIVITY

The methanol extract of *A. caudatus* showed *in vitro* activity against earthworms (*Pheretima posthuma*). Amaranth appears to be a potent worm expeller, causing worm paralysis and death. Both mechanisms of action are better than the only paralytic effect of piperazine in worms. This effect seems to be related to the polyphenolic compounds in the extract (B. Ashok Kumar et al. 2010a).

HEPATOPROTECTIVE FUNCTION

The methanol extract of *A. caudatus* induced favorable changes in serum levels of liver enzymes and oxidative defense markers in a model of paracetamol-induced liver damage in rats (B. S. Ashok Kumar et al. 2011). Further clinical research is required to understand this potential therapeutic application better.

CELIAC DISEASE

The significant proteins of amaranth are albumin (40%), glutelin (25–30%) and globulin (20%). The occurrence of prolamins, a major group of allergens, is minor in amaranth (2–3%). The allergenic potential of prolamins from amaranth was evaluated *in vitro* and *in vivo* with no immune reactivity in intestinal T-cells from celiac disease and transgenic mice (Arendt and Zannini 2013). Further characterization of the protein pattern of *A. caudatus* and the corresponding *in vitro* immunoreactivity with animal anti-gliadin IgG antibodies and human anti-gliadin IgA from celiac patients revealed the safety of amaranth for celiac patients and suggested its inclusion in gluten-free diets (Ballabio et al. 2011).

ANTIDIABETIC ACTIVITY

A. caudatus promotes insulin secretion in murine models (Zambrana et al. 2018; Girija et al. 2011). Accordingly, the glycemic index of a snack containing amaranth, quinoa and tarwi was significantly lower in healthy volunteers (Grados Torrez et al. 2018). Further clinical evaluation is needed to fully understand the antidiabetic properties of amaranth in humans and its mechanism of action.

ANTI-HYPERCHOLESTEROLEMIC AND ANTI-HYPERLIPIDEMIC ACTIVITY

In rabbits, *A. caudatus* extract decreased serum cholesterol, LDL, triglycerides and other cardiovascular risk indicators (Kabiri et al. 2010). This finding is consistent with the presence of tocopherols, lutein and polyunsaturated fatty acids in amaranth. Nevertheless, there is also clinical evidence of the role of amaranth in increasing cardiovascular risk in patients with obesity and who are overweight (Dus-zuchowska et al. 2019). Consequently, further studies are needed to characterize this pharmacological property better.

AMARANTHUS CAUDATUS AS A FUNCTIONAL FOOD

Pseudocereals such as amaranth, quinoa and buckwheat are considered functional food due to their unique nutritional content of essential amino acids and the positive impact of their bioactive compounds on the body's physiological functions (Schmidt et al. 2021).

Carbohydrates constitute the main component of amaranth grains and flour (60–70%), whereas fats are in low percentage (~6%). Amaranth oil possesses a significant concentration of squalene (7–8%). Squalene is one of the principal skin surface lipids exhibiting protection against ultraviolet-induced damage. Available preclinical evidence supports squalene as a promising co-adjuvant in cancer and cholesterol-lowering treatments (Kelly 1999).

Amaranth exhibits a high protein content (~14%). Its amino acid composition is quite close to the daily consumption recommended by the FAO. Moreover, lysine concentration in cereals (i.e., wheat, corn, rice, oat) is limited, while it is remarkably present in amaranth (4.8–6.7%). Among other adverse effects, dietary lysine restriction may induce muscle degradation and lipid accumulation in skeletal muscle (Watanabe et al. 2020; Goda et al. 2021).

The presence of B and E vitamins among the amaranth constituents is also relevant. Niacin promotes endothelial function improvement (Sahebkar 2014). Vitamin E is a potent antioxidant with proven efficacy. In addition, a crucial role of this vitamin has been proposed for cellular signaling, gene regulation, nerve functions and membrane processes (Niki and Abe 2019).

Amaranth consumption may contribute to an optimal fluid balance, muscle contraction and neuronal signaling due to the presence of potassium among its chemical constituents. In addition, it is also an essential source of phosphorus, magnesium and manganese. Phosphorus is responsible for cell and tissue repair and growth.

Magnesium is a cofactor of several enzymes involved in muscle and nerve function, blood glucose control, blood pressure regulation and protein synthesis. Manganese is involved in diverse metabolic routes, including glucose and lipid metabolism, calcium absorption, and contributes to the synthesis and physiology of skeletal, connective and neuronal tissues.

Along with its remarkable nutritional value, as previously mentioned, several studies on amaranth support its anticarcinogenic, antioxidant, anti-inflammatory, antipyretic, antinociceptive, antidepressant, antibacterial, antimalarial, antifungal, anthelmintic, hepatoprotective, antidiabetic, anti-hypercholesterolemic and anti-hyperlipidemic properties. Hence, the regular consumption of amaranth may be associated with wholesome nutrition and optimal health conditions, even in celiac patients.

AMARANTH PROCESSING

Popping

Popping is a technique traditionally applied in Mexico and India for amaranth seeds. When grains are heated to burst without pressure and fat, the water evaporates and the starch granules expand, giving rise to a crunchy, fluffy texture and a savory flavor (Burgos and Armada 2015). The application of this method is only possible for maize kernels, among true cereals.

Due to the brief exposure, the nutrient loss is minimal after this thermal treatment. Moreover, the starch is gelatinized and the proteins are denatured (Bender and Schönlechner 2021). Therefore, popped amaranth can be eaten alone or with syrup, molasses, or milk (Tömösközi et al. 2011).

Milling

The milling process allows access to the high starch content of amaranth grains. In addition, the grains contain a meager degree of moisture (~13%), leading to a good separation between bran and perisperm during the grinding process (Arendt and Zannini 2013). The perisperm can be grounded to obtain flour with excellent nutritional properties (Table 12.1).

Extrusion

Extrusion requires high temperature and pressure to gelatinize the starch and denature protein. The nutrients are maintained within the amaranth grains or flour because of minimal exposure time. The extruded product can be consumed directly or used as a pre-cooked product (Bender and Schönlechner 2021).

FOOD APPLICATIONS

Bakery Products

Amaranth flour appears suitable for preparing unleavened bread. For leavened bread, it contributes as a nutritional enrichment. Nevertheless, gluten absence and its implications in texture, diminished loaf volume and a darker color in baked products are relevant limitations attributed to amaranth flour (Ayo 2007).

Strategies to benefit from this flour's high nutritional quality include replacing flour with a maximum of 10% from amaranth for improving its flavor and nutritional composition (dos Reis Lemos et al. 2022). Moreover, the sensory acceptability of bread made with amaranth and wheat flour is higher due to a characteristic nutty flavor provided by amaranth (Rosentrater and Evers 2018). In addition, the moisture-holding property of amaranth flour increases the shelf-life of this bread (Rosell, Cortez, and Repo-Carrasco 2009).

Gluten-Free Pasta

The individual appropriateness of amaranth, quinoa, and buckwheat flours for pasta production and mixtures of these flours was investigated. The best result obtained in texture and cooking quality corresponds to the combination of 60% buckwheat, 20% amaranth and 20% quinoa (Schoenlechner et al. 2010). A more detailed characterization of gluten-free flours is currently needed to obtain mixtures with an optimal nutritional profile and organoleptic characteristics.

Fermented Food

In Nigeria, "ogi" is a traditional weaning food produced by the fermentation of amaranth grains with lactic acid bacteria (Akingbala and Adeyemi 1994). Fermentation promotes phytates degradation in amaranth flour but to a lesser extent in grains (Castro-Alba et al. 2019). Consequently, it is feasible to achieve a better bioavailability of iron and calcium in products based on fermented amaranth flour.

The use of amaranth flour for sourdough fermentation application is quite promising. Amaranth flour can generate a dough with similar viscosity and elasticity to the traditional wheat-based preparation. Bread prepared with 20% amaranth sourdough and fermented with *Lactobacillus helveticus* presented good sensory acceptability due to the bacterial capacity to release precursor amino acids of compounds that provide aroma to the preparation (Arendt and Zannini 2013).

Beverages

Amaranth is traditionally employed to brew alcoholic and non-alcoholic, fermented and non-fermented beverages. In this context, a high-protein drink similar to skimmed cow's milk composition was developed from amaranth grains. The gluten-free nature and chemical composition of amaranth seeds encourage consumption of this beverage among vegan, lactose-intolerant and celiac patients (Manassero, Añón, and Speroni 2020).

Also, the suitability of amaranth for beer brewing was investigated. A beer prepared from 100% amaranth malt exhibited a better aroma and taste than pure barley malt beer. Nevertheless, reduced foam stability and intensified bitterness were reported (Zweytick and Berghofer 2009).

CONCLUSIONS

Amaranth exhibits a characteristic protein profile and an interesting content of vitamins and minerals. Its gluten-free composition, moisture-binding properties, nutritional contribution and sensory acceptability, among other features, encourage its application in the food industry. The beneficiaries of its properties as a functional food are patients of all age groups, including celiac patients.

In addition, amaranth possesses a wide variety of bioactive compounds. These compounds provide amaranth several prospective nutraceutical applications not only for health maintenance but also for treating diverse diseases.

The functional and nutraceutical benefits of amaranth support the worldwide promotion of its consumption, especially in developing countries, as in the case of Latin America, where it holds an added cultural value.

REFERENCES

Adewale, Adetutu, and Awe Emmanuel Olorunju. 2013. "Modulatory of Effect of Fresh Amaranthus Caudatus and Amaranthus Hybridus Aqueous Leaf Extracts on Detoxify Enzymes and Micronuclei Formation after Exposure to Sodium Arsenite." *Pharmacognosy Research* 5 (4). Pharmacognosy Res: 300–305. http://doi.org/10.4103/0974-8490.118819.

Ahenkora, Kwaku, Stratford Twumasi-Afriyie, Peter Yao, Kanze Sallah, and Kwadwo Obeng-Antwi. 2016. "Protein Nutritional Quality and Consumer Acceptability of Tropical Ghanaian Quality Protein Maize." *Food and Nutrition Bulletin* 20 (3). SAGE PublicationsSage CA: Los Angeles, CA: 354–360. https://doi.org/10.1177/156482659902000313.

Akingbala, J. O., and I. A. Adeyemi. 1994. "Evaluation of Amaranth Grains for Ogi Manufacture." *Plant Foods for Human Nutrition* 46: 19–26.

Al-Dhabi, Naif Abdullah, Mariadhas Valan Arasu, Chang Ha Park, and Sang Un Park. 2015. "An Up-to-Date Review of Rutin and Its Biological and Pharmacological Activities." *EXCLI Journal* 14 (January). Leibniz Research Centre for Working Environment and Human Factors: 59. https://doi.org/10.17179/EXCLI2014-663.

Amico, Stefano D., and Regine Schoenlechner. 2017. "Amaranth: Its Unique Nutritional and Health-Promoting Attributes." In *Gluten-Free Ancient Grains*, 131–159. Texas: Woodhead Publishing Series in Food Science, Technology and Nutrition. https://doi.org/10.1016/B978-0-08-100866-9.00006-6.

Arendt, Elke K., and Emanuele Zannini. 2013. "Amaranth." In *Cereal Grains for the Food and Beverage Industries*, 439–473. Texas: Woodhead Publishing Series in Food Science, Technology and Nutrition. http://doi.org/10.1533/9780857098924.439.

Ashok Kumar, Bagepalli Srinivas, Kuruba Lakshman, Peresandra Avalakondarayppa Arun Kumar, Gollapalle Lakshminarayana Shastry Viswantha, Veeresh Prabhakar Veerapur, Boreddy Shivanadppa Thippeswamy, and Bachappa Manoj. 2011. "Hepatoprotective Activity of Methanol Extract of Amaranthus Caudatus Linn. against Paracetamol-Induced Hepatic Injury in Rats." *Zhong Xi Yi Jie He Xue Bao = Journal of Chinese Integrative Medicine* 9 (2). Zhong Xi Yi Jie He Xue Bao: 194–200. http://doi.org/10.3736/JCIM20110213.

Ashok Kumar, Bagepalli Srinivas, K. Lakshman, K. N. Jayaveera, R. Nandeesh, B. Manoj, and D. Ranganayakulu. 2010a. "Comparative in Vitro Anthelmintic Activity of Three Plants from the Amaranthaceae Family." *Archives of Biological Sciences* 62 (1): 185–189. http://doi.org/10.2298/ABS1001185K.

Ashok Kumar, B. S., K. Lakshman, K. N. Jayaveera, C. Vel Murgan, P. A. Arun Kumar, R. Vinod Kumar, Hegade Meghda, and S. M. Sridhar. 2010b. "Pain Management in Mice Using Methanol Extracts of Three Plants Belongs to Family Amaranthaceae." *Asian Pacific Journal of Tropical Medicine*: 527–530. http://doi.org/10.1016/S1995-7645(10)60127-7.

Ashok Kumar, B. S., K. Lakshman, C. Velmurugan, E. Vishwanath, and S. Gopisetty. 2022. "Evaluation of Antidepressant like Activity in Amaranthus Caudatus." *Mahidol University Journal of Pharmaceutical Sciences* 42 (1): 23–28. https://pharmacy.mahidol.ac.th/journal/_files/2015-42-1_23-28.pdf

Ayo, Jerome Adekunle. 2007. "The Effect of Amaranth Grain Flour on the Quality of Bread." *International Journal of Food Properties* 4 (2). http://doi.org/10.1081/JFP-100105198. Taylor & Francis Group: 341–351. http://doi.org/10.1081/JFP-100105198.

Ballabio, Cinzia, Francesca Uberti, Chiara di Lorenzo, Andrea Brandolini, Elena Penas, and Patrizia Restani. 2011. "Biochemical and Immunochemical Characterization of Different Varieties of Amaranth (Amaranthus L. Ssp.) as a Safe Ingredient for Gluten-Free Products." *Journal of Agricultural and Food Chemistry* 59: 12969–12974. http://doi.org/10.1021/jf2041824.

Bender, D., and R. Schönlechner. 2021. "Recent Developments and Knowledge in Pseudocereals Including Technological Aspects." *Acta Alimentaria* 50 (4). Akadémiai Kiadó: 583–609. http://doi.org/10.1556/066.2021.00136.

Broekaert, Willem F., Wim Mariën, Franky R. G. Terras, Miguel F. C. De Bolle, Jozef Vanderleyden, Bruno P. A. Cammue, Paul Proost, et al. 1992. "Antimicrobial Peptides from Amaranthus Caudatus Seeds with Sequence Homology to the Cysteine/Glycine-Rich Domain of Chitin-Binding Proteins." *Biochemistry* 31 (17). Biochemistry: 4308–4314. http://doi.org/10.1021/BI00132A023.

Burgos, Verónica Elizabeth, and Margarita Armada. 2015. "Characterization and Nutritional Value of Precooked Products of Kiwicha Grains (*Amaranthus Caudatus*)." *Food Science and Technology* 35 (3). Sociedade Brasileira de Ciência e Tecnologia de Alimentos: 531–538. http://doi.org/10.1590/1678-457X.6767.

Castro-Alba, Vanesa, Claudia E. Lazarte, Daysi Perez-Rea, Nils Gunnar Carlsson, Annette Almgren, Björn Bergenståhl, and Yvonne Granfeldt. 2019. "Fermentation of Pseudocereals Quinoa, Canihua, and Amaranth to Improve Mineral Accessibility through Degradation of Phytate." *Journal of the Science of Food and Agriculture* 99 (11). Wiley-Blackwell: 5239–5248. http://doi.org/10.1002/JSFA.9793.

Dus-zuchowska, Monika, Jaroslaw Walkowiak, Anna Morawska, Patrycja Krzyzanowska-Jankowska, Anna Miskiewicz-chotnicka, Juliusz Przyslawski, and Aleksandra Lisowska. 2019. "Amaranth Oil Increases Total and LDL Cholesterol Levels without Influencing Early Markers of Atherosclerosis in an Overweight and Obese Population: A Randomized Double-Blind Cross-Over Study in Comparison with Rapeseed Oil Supplementation." *Nutrients* 11 (12). Multidisciplinary Digital Publishing Institute: 3069. http://doi.org/10.3390/NU11123069.

Emekli, Nefise Yasemin, Berna Kendirli, and Ahmet Kurunc. 2012. "Phytochemical Constituents and Antimicrobial Activity of Leaf Extracts of Three Amaranthus Plant Species." *African Journal of Biotechnology* 9 (21): 3178–3182. http://doi.org/10.4314/ajb.v9i21.

Ferreira, Tania Aparecida, and Jose Alfredo Gomes Arêas. 2010. "Calcium Bioavailability of Raw and Extruded Amaranth Grains." *Food Science and Technology* 30 (2). Sociedade Brasileira de Ciência e Tecnologia de Alimentos: 532–538. http://doi.org/10.1590/S0101-20612010000200037.

Food and Agriculture Organization, World Health Organization, and United Nations University. 2007. *Protein and Amino Acid Requirements in Human Nutrition WHO Technical Report Series 935*. Geneva: WHO Library Cataloguing-in-Publication Data. www.who.int/bookorders.

Girija, K., K. Lakshman, Chandrika Udaya, Ghosh Sabhya Sachi, and T. Divya. 2011. "Antidiabetic and Anti-Cholesterolemic Activity of Methanol Extracts of Three Species of Amaranthus." *Asian Pacific Journal of Tropical Biomedicine* 1 (2). China Humanity Technology Publishing House: 133. http://doi.org/10.1016/S2221-1691(11)60011-7.

Goda, Yuki, Daisuke Yamanaka, Hiroki Nishi, Masato Masuda, Hiroyasu Kamei, Mikako Kumano, Koichi Ito, et al. 2021. "Dietary Lysine Restriction Induces Lipid Accumulation in Skeletal Muscle through an Increase in Serum Threonine Levels in Rats." *Journal of Biological Chemistry* 297 (4). Elsevier: 1–12. http://doi.org/10.1016/J.JBC.2021.101179.

Grados, Torrez, Ricardo Enrique, Rodrigo Daniel Trino, Julio Pérez Gonzáles, and Eduardo Gonzáles Dávalos. 2018. "Determinación Del Índice Glucémico de Un Producto Elaborado a Base de Amaranto (Amaranthus Caudatus Linnaeus), Quinua (Chenopodium Quinoa Willd) y Tarwi (Lupinus Mutabilis Sweet) Para Tratamiento Coadyuvante de Diabetes Tipo 2 y Obesidad." *Revista CON-CIENCIA* 6 (1). Facultad de Ciencias Farmacéuticas y Bioquímicas: 73–82. www.scielo.org.bo/scielo.php?script=sci_arttext&pid=S2310-02652018000100008&lng=es&nrm=iso&tlng=es.

Herrera, Fontana, María Elisa, Aida Maribel Chisaguano Tonato, Jessica Verónica Jumbo Crisanto, Nancy Pepita Castro Morillo, and Andrea Paola Anchundia Ortega. 2020. *Tabla de Composición Química de Los Alimentos: Basada En Nutrientes de Interés Para La Población Ecuatoriana | Bitácora Académica*. Quito: USFQ Press. https://revistas.usfq.edu.ec/index.php/bitacora/issue/view/191/PDF%20Bit%C3%A1cora%20Acad%C3%A9mica%20Vol.%2011.

Juliano, B. O. 2016. "Rice: Role in Diet." In *Encyclopedia of Food and Health*, 641–645. Oxford: Academic Press. http://doi.org/10.1016/B978-0-12-384947-2.00595-X.

Kabiri, Najmeh, Seddigheh Asgary, Hossein Madani, and Parvin Mahzouni. 2010. "Effects of Amaranthus Caudatus l. Extract and Lovastatin on Atherosclerosis in Hypercholesterolemic Rabbits." *Journal of Medicinal Plants Research* 4 (5): 355–361. www.academicjournals.org/JMPR.

Karamać, Magdalena, Francesco Gai, Erica Longato, Giorgia Meineri, Michał A. Janiak, Ryszard Amarowicz, and Pier Giorgio Peiretti. 2019. "Antioxidant Activity and Phenolic Composition of Amaranth (Amaranthus Caudatus) during Plant Growth." *Antioxidants* 8 (6). Multidisciplinary Digital Publishing Institute (MDPI). http://doi.org/10.3390/ANTIOX8060173.

Kelly, G. S. 1999. "Squalene and Its Potential Clinical Uses." *Alternative Medicine Review* 4 (1): 29–36. https://pubmed.ncbi.nlm.nih.gov/9988781/.

Luengo Fernández, Emilio, ed. 2007. *Alimentos Funcionales y Nutracéuticos*. Madrid: Sociedad Española de Cardiología.

Manassero, Carlos Alberto, María Cristina Añón, and Francisco Speroni. 2020. "Development of a High Protein Beverage Based on Amaranth." *Plant Foods for Human Nutrition (Dordrecht, Netherlands)* 75 (4). Plant Foods Hum Nutr: 599–607. http://doi.org/10.1007/S11130-020-00853-9.

Martinez-Lopez, Alicia, Maria C. Millan-Linares, Noelia M. Rodriguez-Martin, Francisco Millan, and Sergio Montserrat-de la Paz. 2020. "Nutraceutical Value of Kiwicha (Amaranthus Caudatus L.)." *Journal of Functional Foods* 65 (February). Elsevier: 103735. http://doi.org/10.1016/J.JFF.2019.103735.

Martínez-Villaluenga, Cristina, Elena Peñas, and Blanca Hernández-Ledesma. 2020. "Pseudocereal Grains: Nutritional Value, Health Benefits and Current Applications for the Development of Gluten-Free Foods." *Food and Chemical Toxicology*: 111178. http://doi.org/10.1016/j.fct.2020.111178.

Molina Vélez, Magdalena. 2008. *Fitoterapia*. Cuenca: Casa de la Cultura Ecuatoriana "Benjamín Carrión" Núcleo del Azuay.

National Research Council. 1984. *Amaranth: Modern Prospects for an Ancient Crop. Amaranth*. Washington, DC: National Academies Press. http://doi.org/10.17226/19381.

Niki, Etsuo, and Kouichi Abe. 2019. "Vitamin E: Structure, Properties and Functions." *Food Chemistry, Function and Analysis* 2019-January (11). Royal Society of Chemistry: 1–11. http://doi.org/10.1039/9781788016216-00001.

Obidike, I. C., B. Amodu, and M. O. Emeje. 2015. "Antimalarial Properties of SAABMAL (®): An Ethnomedicinal Polyherbal Formulation for the Treatment of Uncomplicated Malaria Infection in the Tropics." *The Indian Journal of Medical Research* 141 (2). Indian J Med Res: 221–227. http://doi.org/10.4103/0971-5916.155585.

Peter, Kavita, and Puneet Gandhi. 2017. "Rediscovering the Therapeutic Potential of Amaranthus Species: A Review." *Egyptian Journal of Basic and Applied Sciences* 4 (3). No Longer Published by Elsevier: 196–205. http://doi.org/10.1016/J.EJBAS.2017.05.001.

Pomeranz, Y., V. L. Youngs, and G. S. Robbins. 1973. "Protein Content and Amino Acid Composition of Oat Species and Tissues." *Cereal Chemistry* 50: 702–707. www.cerealsgrains.org/publications/cc/backissues/1973/Documents/chem50_702.pdf.

Reis Lemos, Andréa dos, Vanessa Dias Capriles, Maria Elisabeth Machado Pinto Silva, and Gomes Areas José Alfredo. 2022. "Effect of Incorporation of Amaranth on the Physical Properties and Nutritional Value of Cheese Bread." *Ciencia y tecnología de alimentos* 32 (3): 427–431. Accessed January 26. http://doi.org/10.1590/S0101-20612012005000079.

Ríos, Montserrat, M. J. Koziol, H. Borgtoft Pedersen, and G. Granda, eds. 2007. *Plantas Útiles Del Ecuador: Aplicaciones, Retos y Perspectivas*. Quito: Abya-Yala.

Rosell, Cristina M., Gladys Cortez, and Ritva Repo-Carrasco. 2009. "Breadmaking Use of Andean Crops Quinoa, Kañiwa, Kiwicha, and Tarwi." *Cereal Chemistry* 86 (4). John Wiley & Sons, Ltd: 386–392. http://doi.org/10.1094/CCHEM-86-4-0386.

Rosentrater, Kurt A., and A. D. Evers. 2018. "Introduction to Cereals and Pseudocereals and Their Production." In *Kent's Technology of Cereals*, 1–76. Cambridge: Woodhead Publishing. http://doi.org/10.1016/B978-0-08-100529-3.00001-3.

Sahebkar, Amirhossein. 2014. "Effect of Niacin on Endothelial Function: A Systematic Review and Meta-Analysis of Randomized Controlled Trials." *Vascular Medicine (United Kingdom)* 19 (1). SAGE PublicationsSage UK: London, England: 54–66. http://doi.org/10.1177/1358863X13515766.

Schmidt, Davi, Marta Regina Verruma-Bernardi, Victor Augusto Forti, and Maria Teresa Mendes Ribeiro Borges. 2021. "Quinoa and Amaranth as Functional Foods: A Review." *Food Reviews International*. Taylor & Francis. http://doi.org/10.1080/87559129.2021.1950175.

Schoenlechner, Regine, Julian Drausinger, Veronika Ottenschlaeger, Katerina Jurackova, and Emmerich Berghofer. 2010. "Functional Properties of Gluten-Free Pasta Produced from Amaranth, Quinoa and Buckwheat." *Plant Foods for Human Nutrition (Dordrecht, Netherlands)* 65 (4). Plant Foods Hum Nutr: 339–349. http://doi.org/10.1007/S11130-010-0194-0.

Shoup, F. K., Y. Pomeranz, and C. W. Deyoe. 1966. "Amino Acid Composition of Wheat Varieties and Flours Varying Widely in Bread-Making Potentialities." *Journal of Food Science* 31 (1). John Wiley & Sons, Ltd: 94–101. http://doi.org/10.1111/J.1365-2621.1966.TB15420.X.

Soriano-García, Manuel, and Isabel Saraid Aguirre-Díaz. 2020. "Nutritional Functional Value and Therapeutic Utilization of Amaranth." In *Nutritional Value of Amaranth*, edited by Viduranga Y. Waisundara. London: IntechOpen. http://doi.org/10.5772/INTECHOPEN.86897.

Srinivas, Bagepalli, Ashok Kumar, Kuruba Lakshman, Korala Konta, Narsimha Jayaveera, Devangam Sheshadri Shekar, Chinna, and Swamyvel Muragan. 2010. "Antinociceptive and Antipyretic Activities of Methanol Extract Amaranthus Caudatus Linn." *Latin American Journal of Pharmacy* 29 (4): 635–639.

Tang, Yao, and Rong Tsao. 2017. "Phytochemicals in Quinoa and Amaranth Grains and Their Antioxidant, Anti-Inflammatory, and Potential Health Beneficial Effects: A Review." *Molecular Nutrition & Food Research* 61 (7). Mol Nutr Food Res. http://doi.org/10.1002/MNFR.201600767.

Tömösközi, Sándor, Lilla Gyenge, Ágnes Pelcéder, Tibor Abonyi, Regine Schönlechner, and Radomir Lásztity. 2011. "Effects of Flour and Protein Preparations from Amaranth and Quinoa Seeds on the Rheological Properties of Wheat-Flour Dough and Bread Crumb." *Czech Journal of Food Sciences* 29 (2): 109–116.

Vega-Gálvez, Antonio, Margarita Miranda, Judith Vergara, Elsa Uribe, Luis Puente, and Enrique A. Martínez. 2010. "Nutrition Facts and Functional Potential of Quinoa (Chenopodium Quinoa Willd.), an Ancient Andean Grain: A Review." *Journal of the Science of Food and Agriculture* 90 (15): 2541–2547. doi:10.1002/JSFA.4158.

Watanabe, Genya, Hiroyuki Kobayashi, Masahiro Shibata, Masatoshi Kubota, Motoni Kadowaki, and Shinobu Fujimura. 2020. "Reduction in Dietary Lysine Increases Muscle Free Amino Acids through Changes in Protein Metabolism in Chickens." *Poultry Science* 99 (6). Elsevier: 3102–3110. http://doi.org/10.1016/J.PSJ.2019.11.025.

Youssef, M. K. E., A. G. Nassar, F. A. El-Fishawy, and M. A. Mostafa. 2016. "Assessment of Proximate Chemical Composition and Nutritional Status of Wheat Biscuits Fortified with Oat Powder." *Assiut Journal of Agricultural Sciences* 47 (5): 83–94. www.aun.edu.eg/faculty_agriculture.

Zambrana, Silvia, Lena C. E. Lundqvist, Virginia Veliz, Sergiu Bogdan Catrina, Eduardo Gonzales, and Claes Göran Östenson. 2018. "Amaranthus Caudatus Stimulates Insulin Secretion in Goto-Kakizaki Rats, a Model of Diabetes Mellitus Type 2." *Nutrients* 10 (1). Multidisciplinary Digital Publishing Institute: 94. http://doi.org/10.3390/NU10010094.

Zweytick, Gernot, and Emmerich Berghofer. 2009. "Production of Gluten-Free Beer." In *Gluten-Free Food Science and Technology*, edited by Eimear Gallagher, 181–199. Singapore: John Wiley & Sons.

13 *Cinchona pubescens* (Cascarilla), Native Species from the Andes Mountains of South America

Juan Peñarreta, Patricia Manzano and Efrén Santos

CONTENTS

BOTANICAL DESCRIPTION

Cinchona (also known as Cascarilla or Quina) is the most commercially important genus of the family *Rubiaceae* (coffee family) after the genus Coffea, which produces the commercial coffee. The genus *Cinchona* was named after the Countess of Chinchón, wife of the Viceroy of Peru, by the Swedish botanist Linnaeus in 1742. According to the well cited legend, the countess was cured of malaria, having administered the bark of *Cinchona* in 1638 after all other remedies failed. Although this story is not true, *Cinchona* ever since was frequently used as a malaria remedy, especially distributed by the Jesuits in their world travels. *Cinchona* is the national tree of Ecuador and is on the coat of arms of Peru (Jäger 2014)

Currently, *Cinchona* (Family: Rubiaceae), is a genus reduced to 24 species distributed in Costa Rica, Panama, the *Cordillera de la Costa* in Venezuela, and in the Andes, from Venezuela to central Bolivia; absent in Brazil, the Guianas, México and the Southern Cone (Ulloa et al. 2017). The native range extends from Andean South America in Bolivia north to Costa Rica. Unusually, it occurs on both sides of the Cordillera in sub-montane rainforest, predominately at altitudes of 800–2,800 m but can be found up to 3500 m (CABI 2022).

DOI: 10.1201/9781003173991-16

TABLE 13.1
Species of the Genus *Cinchona* Currently Accepted, with Their Respective Geographical Distribution (Gerardo and Aymard 2019)

	Countries/Regions					
Species	Bolivia	America central	Colombia	Ecuador	Peru	Venezuela
Cinchona antioquiae L. Andersson (1998)			X			
Cinchona anderssonii C. D. Maldonado et al. (2017)	X					
Cinchona asperifolia Wedd. (1848)	X					
Cinchona barbacoensis H. Karst. (1859)			X	X		
Cinchona calisaya Wedd. (1948)	X	X	X		X	
Cinchona capuli L. Andersson (1994)			X	X	X	
Cinchona carabayensis Wedd. (1848) = *C. calisaya* Wedd.	X				X	
Cinchona fruticosa L. Andersson (1998)				X	X	
Cinchona glandulifera (Ruiz) Ruiz & Pav. (1802)					X	
Cinchona govana Miq. (1861) = *C. pubescens* Vahl	X				X	
Cinchona hirsuta Ruiz & Pav. (1799)				X	X	
Cinchona lancifolia Mutis (1793)			X	X		X
Cinchona lucumifolia Pav. ex Lindl. (1838)				X		
Cinchona macrocalyx Pav. ex DC. (1829)	X			X	X	
Cinchona micrantha Ruiz & Pav. (1799)	X				X	
Cinchona mutisii Lamb. (1821)				X	X	
Cinchona nitida Ruiz & Pav. (1799)					X	
Cinchona officinalis L. (1753)	X		X	X	X	
Cinchona parabolica Pav. (1859)				X	X	
Cinchona pitayensis (Wedd.) Wedd. (1849)			X	X	X	
Cinchona pubescens Vahl (1790)	X	X	X	X	X	X

Species	Countries/Regions					
	Bolivia	America central	Colombia	Ecuador	Peru	Venezuela
Cinchona pyrifolia L. Andersson (1998)					X	
Cinchona scrobiculata Bonpl. (1808)				X	X	
Cinchona villosa Pav. ex Lindl. (1838)	X			X	X	

FIGURE 13.1 *C. pubescens* trees in the natural range: Isla Santa Cruz, Parroquia Santa Rosa, Cantón Santa Cruz, Galápagos, Ecuador (-0.62882–90.36388). (Photos by Francisco León.)

The last species described to date is *C. anderssonii* Maldonado, a small tree shrub, presently known only in montane forests (2,200–2,600 m) of the Yungas region of Bolivia (Maldonado et al. 2017).

The trees are usually 4–10 m tall (Figure 13.1). The leaves are broadly elliptic-ovate or sometimes sub-orbicular, of 10–20 cm long and 7–10.5 cm wide; with upper surface puberulent, sometimes primarily along veins, or glabrate, with lateral veins usually 9–11 pairs, margins entire, apex rounded, base broadly to narrowly cuneate, with petioles of 1.5–4.5 cm long, stipules ovate, and caducous (Wagner, Herbst, and Sohmer 1999). In Ecuador, around Loja, *Cinchona* flowers and bears fruits at the same time all year round (Gray, Arrot, and Miller 1737). The flowers of *C. pubescens* are clustered in large, broadly pyramidal panicles, usually up to 20 cm but sometimes longer in size. The corollas are pinkish or purplish, paler at base (corollas outside may be white to light pink or red in Hawaii and Galápagos) and are fragrant (Andersson 1998; Starr, Starr, and Loope 2003). It was thus assumed that *C. pubescens* is insect-pollinated but this has not been confirmed anywhere (Starr, Starr, and Loope 2003).

The corolla tube is 9–14 mm long, pubescent outside and glabrous inside. The capsules are ellipsoid to subcylindrical and 13–41 × 5–7 mm long, opening from the base to tip when mature. Seeds are 7–12 × 2.1–2.8 mm, including the irregularly dentate wings (Jäger 2014).

TAXONOMY AND HYBRIDIZATION

The taxonomy of *C. pubescens* is especially difficult since it frequently hybridizes with other *Cinchona* species where they occur together in nature (Camp 1949; Andersson and Taylor 1994; Jäger 2014). Further studies could be performed for species identification and phylogenetic analysis including DNA barcoding. In 1946, 4 species were known to form hybrids with *C. pubescens* (Acosta Solís 1945), whereas Andersson (1998) lists 7 species hybridizing with it (*C. barbacoensis, C. calisaya, C. lancifolia, C. lucumifolia, C. macrocalyx, C. micrantha, C. officinalis*). Hybrids between *C. pubescens* and *C. calisaya* are the most commonly found in nature and seem also to have been produced in cultivation (Andersson 1998).

TRADITIONAL USES

Locally in Ecuador, the bark is used to extract quinine, the alkaloid used to cure malaria and fevers. This species is abundant in quinine and other alkaloids (unspecified ethnic group—from provinces of Chimborazo, Bolívar, Azuay, Cañar, and Napo, Ecuador). Quinine is used to treat heart conditions; it is a tonic, euptic and anti-fermentative in chronic stomach colds with acidic fermentation (promotes digestion) (unspecified ethnic group—province of Bolívar, Ecuador). The bark, mixed with brandy, is used to treat colds and sore throat (unspecified ethnic group—province of Loja, Ecuador) (de la Torre et al. 2008). Quinine sulfate is extracted from the bark, which is used in the manufacture of contraceptive condoms (unspecified ethnic group—province of Bolívar, Ecuador) (Paniagua-Zambrana, Bussmann, and Romero 2020). On the other hand, the Cascarilla water extracted from the crust is used to combat hair loss (de la Torre et al. 2008).

In Colombia, stems and roots are used as an analgesic; the bark is used to treat diarrhea, fever, indigestion and malaria. The bark of Cascarilla has healing, anti-inflammatory and antihemorrhagic properties; it is also febrifuge, it heals wounds and sores, and it is antimalarial, antidiarrheal and facilitates digestion. Cascarilla is also used in the treatment of inappetence, cardiac arrhythmias, gastroduodenal ulcers and inflammations of the small intestine (Paniagua-Zambrana, Bussmann, and Romero 2020).

MAIN CONSTITUENTS

Among the many thousands of natural products isolated and characterized so far, *Cinchona* alkaloids with quinine as a major member occupy an exceptional position in human civilization (Toovey 2004). In Ecuador, *C. pubescens* is naturally grown in the province of Bolívar and on the western slopes of the Andes, producing the highest alkaloid content. However, the alkaloid content varies considerably with locality, from 0.1 to 7.8%, and with the age of the tree and the tissue. Younger trees and thicker barks contain more alkaloids (Jäger, 2011).

Pelletier & Carentouaislaz were the first to isolate the alkaloids from *C. pubescens*, the most prominent being the presence of quinine and quinidine stereoisomer. Other compounds that have been isolated are cinonine, cinchonidine and a variety of quinoline derivatives. Alkaloids are undoubtedly the main component (approximately 6.5% of total alkaloids, approximately 20) among which the majority is quinine representing 70–90% of the pair of stereoisomers, 1% corresponds to quinidine, along with its 6-dimethoxy analogs as cinchonine and cinchonidine, which are useful as antimalarial and used entirely for the preparation of quina. Another group of compounds found in *C. pubescens* are the anthraquinones. Essential oils are also an important group and represent 0.02 to 0.08% of its composition (Noriega et al. 2015).

A variety of a stringent components have been identified (dimers and trimers proantocianidoles tannins, catechin, tannins 8% of total tannins) and other compounds such as flavonoids, catechin, kaempferol, apigenin and quercetin; glycosides, organic acids (quinotanic acid, cinconic red), monoglycosides such as quinovic acid (3β-hydroxyurea-dihydroxy benzoic acid) and terpene compounds (Aerts et al. 2007).

PHARMACOLOGICAL ACTIVITY

A most important activity of *Cinchona* alkaloids is the antiprotozoal action of quinine, used for more than 400 years for the treatment of malaria (Casteel 1997; Toovey 2004). Malaria is one of the life-threatening infections caused by a protozoan parasite. It is still a major public health concern of most endemic areas of the world. Five human *Plasmodium* species (*Plasmodium falciparum, P. vivax, P. ovale, P. knowlesi*, and *P. malariae*) cause malaria infection. The major complications are caused by *P. falciparum* and *P. vivax*, with *P. falciparum* being the more virulent. It is indicated that about 1–3 million mortalities per year, mainly in children and pregnant women, are due to severe malaria caused by *P. falciparum*. These pathologies are severe anemia, cerebral malaria and acute respiratory distress (Geleta and Ketema 2016). The cardiological effects of *Cinchona* bark alkaloids have been recognized in academic medicine at the end of the 17th century. Quinine was used at the beginning, but its pseudoenantiomer quinidine has been found to have more beneficial antiarrhythmic properties, thus becoming a standard medication until newer drugs were developed in the mid-20th century (Klevans, Kelly, and Kovacs 1977).

The major problem with the use of this alkaloid is its quick absorption by the gastrointestinal tract giving risk of overdosing. This may lead to diastolic arrest (ventricular arrhythmia) and death. Safety concerns resulted in the significant reduction of quinidine use in a therapy over years. However, recently renewed interest of the medical use of quinidine is observed; in particular, quinidine combination with verapamil has been reported to be safe and efficacious in the treatment of atrial fibrillation. Quinidine has also been used successfully to treat idiopathic ventricular fibrillation, Brugada and short QT syndromes. Quinidine applications in modern cardiology are the subject of reviews (Grace and Camm 1998; Yang et al. 2009).

Muscle cramp is a recurrent and painful condition and a common complaint among elderly subjects and patients treated with hemodialysis. Meta-analysis of the different 23 trials with a total of 1,586 participants has been published with conclusion that quinine at the dosage of typically 300 mg/day (range 200–500 mg), compared to placebo, significantly reduced cramp numbers over 2 weeks by 28%, cramp intensity by 10%, and cramp days by 20%, but without affecting the cramp duration (Kacprzak 2015).

On the other hand, the antioxidant activity of the dry extracts from *Cinchona pubescens* Vahl species was determined by spectrophotometric methods ABTS and 1-diphenyl-2-picrylhydrazyl (DPPH), expressed as IC50 (50% inhibition of radical oxidation) were 42.00 ± 0.2 μg/ml (DPPH) and 88.00 ± 0.2 IC50 μg/ml ABTS. In cosmetic formulations (gels and creams) evaluated through photochemiluminescence, a significantly increased antioxidant potential was observed when compared to the reference formula. The potential was up to 15 times when compared to a dry extract of *Camellia sinensis* (Noriega et al. 2015).

CONCLUSION

Cinchona pubescens is an important component of traditional medicine in different countries of Central and South America including Ecuador. For centuries, *C. pubescencs* have been used to treat different illnesses including malaria; and the quinine extracted from the bark is used as a tonic, eupticm and anti-fermentative in chronic stomach colds. Furthermore, *C. pubescens* exhibit various pharmacological activities including antiprotozoal, antiarrhythmic, antispasmodic and antioxidant. Therefore, further studies need to be implemented to obtain more information on their mechanism of action, toxicity and chemical applications for the development of new drugs.

REFERENCES

Acosta Solís, M. 1945. "Botánica de Las Cinchonas." *Revista Al Servicio de Las Ciencias Naturales y Biológicas. Instituto Ecuatoriano de Ciencias Naturales* 15/16: 29–55.

Aerts, Rob J., Anthony De Waal, Ed J. M. Pennings, and Rob Verpoorte. 2007. "The Distribution of Strictosidine-Synthase Activity and Alkaloids in Cinchona Plants." *Planta* 183 (4): 536–541. https://doi.org/10.1007/BF00194275.

Andersson, L. 1998. "A Revision of the Genus Cinchona (Rubiaceae-Cinchoneae)." *Memoirs of the New York Botanic Garden* 80.

Andersson, L., and C. M. Taylor. 1994. *Rubiaceae, Cinchoneae, Coptosapelteae*. Edited by G. Harling and L. Andersson. Flora of Ecuador. Council for Nordic Publications in Botany. https://books.google.com.ec/books?id=H5cgkgAACAAJ.

CABI. 2022. *Cinchona pubescens* (quinine tree). Compendio de Especies Invasoras. www.cabi.org/isc

Camp, W. H. 1949. "Cinchona at High Altitudes in Ecuador." *Brittonia* 6 (4): 394–430. https://doi.org/10.2307/2804925.

Casteel, D. A. 1997. *Antimalarial Agents. Burger's Medicinal Chemistry and Drug Discovery, Vol 5*. Wolff ME. New York: Wiley.

Geleta, Getachew, and Tsige Ketema. 2016. "Severe Malaria Associated with *Plasmodium Falciparum and P. Vivax* among Children in Pawe Hospital, Northwest Ethiopia." *Malaria Research and Treatment* 2016 (March): 1–7. https://doi.org/10.1155/2016/1240962.

Gerardo, A., and C. Aymard. 2019. "Breve Reseña de Los Aspectos Taxonómicos y Nomenclaturales Actuales Del Género Cinchona (Rubiaceae- Cinchoneae) A Brief Outline on Current Taxonomical and Nomenclatural Aspects of the Genus Cinchona (Rubiaceae- Cinchoneae)." *Revista de la Academia Colombiana de Ciencias Exactas Físicas y Naturales* 43: 234–241. http://doi.org/10.18257/raccefyn.1079.

Grace, A. A., and A. J. Camm. 1998. "Quinidine." *The New England Journal of Medicine* 338 (1): 35–45. https://doi.org/10.1056/NEJM199801013380107.

Gray, J., W. Arrot, and P. Miller. 1737. *An Account of the Peruvian or Jesuit's Bark*. 40th ed. Philosophical YTransactions (1683–1775).

Jäger, H. 2011. Cinchona pubescens. In: Roloff, A., Weisgerber, H., Lang, U., Stimm, B. (Eds.), *Enzyklopädie der Holzgewächse*, Weinheim: Wiley-VCH, 58. Erg.Lfg. 06/11, 14 pp.

Kacprzak, Karol. 2015. "Chemistry and Biology of Cinchona Alkaloids." *Natural Products*, January 2013. https://doi.org/10.1007/978-3-642-22144-6.

Klevans, L. R., R. J. Kelly, and J. L. Kovacs. 1977. "Comparison of the Antiarrhythmic Activity of Quinidine and Quinine." *Archives Internationales de Pharmacodynamie et de Therapie* 227 (1): 57–68. http://europepmc.org/abstract/MED/901074.

la Torre, L. de, H. Navarrete, P. Muriel, M. J. Macía, and H. Balslev. 2008. *Enciclopedia de Las Plantas Útiles Del Ecuador*. Edited by L. de la Torre, H. Navarrete, P. Muriel, M. J. Macía, and H. Balslev. Quito, Ecuador: Quito/Aarhus: Herbario QCA de la Escuela de Ciencias Biológicas de la Pontificia Universidad Católica del Ecuador/Herbario AAU del Departamento de Ciencias Biológicas de la Universidad de Aarhus.

Maldonado, Carla, Claes Persson, Joaquina Alban, Alexandre Antonelli, and Nina Rønsted. 2017. "*Cinchona Anderssonii* (Rubiaceae), a New Overlooked Species from Bolivia." *Phytotaxa* 297 (2): 203–208. http://doi.org/10.11646/phytotaxa.297.2.8.

Noriega, Paco, María Sola, Angelka Barukcic, Katic Garcia, and Edison Osorio. 2015. "Cosmetic Antioxidant Potential of Extracts from Species of the *Cinchona Pubescens* (Vahl)." *International Journal of Phytocosmetics and Natural Ingredients* 2 (1): 14. https://doi.org/10.15171/ijpni.2015.14.

Paniagua-Zambrana, N. Y., Bussmann, R. W., and Romero, C. 2020. *Cinchona officinalis L.Cinchona pubescens* VahlRubiaceae. Ethnobotany of the Andes, 1–6. doi:10.1007/978-3-319-77093-2_72-1.

Starr, Forest, Kim Starr, and Lloyd Loope. 2003. "*Cinchona Pubescens*." Hawaii. www.hear.org/starr/hiplants/reports/pdf/cinchona_pubescens.pdf.

Toovey, S. 2004. *The Miraculous Fever-Tree. The Cure That Changed the World Fiametta Rocco*. Edited by S Toovey. San Francisco: Harper Collins.

Ulloa, Carmen Ulloa, Pedro Acevedo-rodríguez, Stephan Beck, Manuel J. Belgrano, Rodrigo Bernal, Paul E. Berry, Lois Brako, et al. 2017. "An Integrated Assessment of the Vascular Plant Species of the Americas." *Science* 358 (6370) (December): 1614–1617. https://doi.org/10.1126/science.aao0398.

Wagner, W. L., D. R. Herbst, and S. H. Sohmer. 1999. *Manual of the Flowering Plants of Hawaii*. Edited by University of Hawaii and Bishop Museum Press. Honolulu, HI: Bishop Museum Special Publication 83.

Yang, Felix, Sam Hanon, Patrick Lam, and Paul Schweitzer. 2009. "Quinidine Revisited." *The American Journal of Medicine* 122 (4): 317–321. https://doi.org/10.1016/j.amjmed.2008.11.019.

Index

Page numbers in *italics* indicate figures; page numbers in **bold** indicate tables.

D

E

O

P

Q

R

S

T

U

V

W

X